# Collins

**NEW GCSE SCIENCE**

# Science B
## Science in Context
# AQA

**Series Editor: Ken Gadd**

**Author team: 4science**

**Student Book**

William Collins' dream of knowledge for all began with the publication of his first book in 1819. A self-educated mill worker, he not only enriched millions of lives, but also founded a flourishing publishing house. Today, staying true to this spirit, Collins books are packed with inspiration, innovation and practical expertise. They place you at the centre of a world of possibility and give you exactly what you need to explore it.

Collins. Freedom to teach

Published by Collins
An imprint of HarperCollins*Publishers*
77 – 85 Fulham Palace Road
Hammersmith
London
W6 8JB

Browse the complete Collins catalogue at
www.collinseducation.com

© HarperCollins*Publishers* Limited 2011

10 9 8 7 6 5 4 3 2 1

ISBN-13 978 0 00 741466 6

British Library Cataloguing in Publication Data
A Catalogue record for this publication is available from the British Library


Commissioned by Letitia Luff
Project managed by Hanneke Remsing and 4science
Packaged by 4science
Proofread by Life Lines Editorial Services
Indexed by Nigel d'Auvergne
New illlustrations by 4science
Concept design by Anna Plucinska
Picture research by Caroline Green
Cover design by Julie Martin
Production by Kerry Howie
Contributing authors John Beeby and Ed Walsh
'Bad Science' pages based on the work of Ben Goldacre

Printed and bound by L.E.G.O. S.p.A. Italy

Acknowledgements – see page 288

# Contents

## ◉ Unit 1: My world

# Contents

## Unit 2: My family and home

# Unit 3: Making my world a better place

# How to use this book

## Welcome to Collins New GCSE Science for AQA Science B!

### The main content

Each two-page lesson has three levels:

> The first part outlines a basic scientific idea.

> The second part builds on the basics and develops the concept.

> The third part extends the concept or challenges you to apply it in a new way.

Information that is only relevant to the Higher tier is indicated with 'Higher tier'.

Each part contains a set of level-appropriate questions that allow you to check and apply your knowledge.

Look for:

> 'Essential notes' boxes

> Internet search terms (at the bottom of every page)

> 'Did you know' and 'Remember' boxes

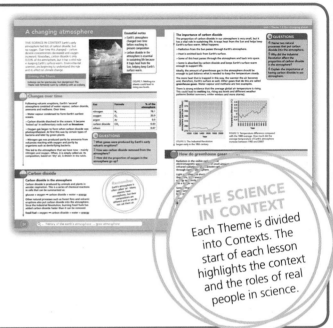

**THE SCIENCE IN CONTEXT**

Each Theme is divided into Contexts. The start of each lesson highlights the context and the roles of real people in science.

### Units and Themes

Each Unit is divided into Themes.

At the start of each Theme, link the science that you will learn with your existing scientific knowledge.

### Checklists

Each Theme contains a checklist.

Summarise the key ideas that you have learned so far and see what you need to know to progress.

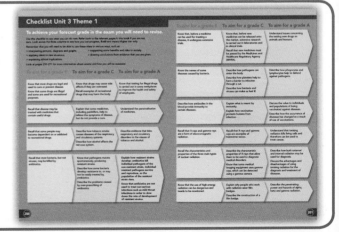

## Exam-style questions

Each Unit contains practice exam-style questions for both Foundation and Higher tiers, labelled with the Assessment Objectives that they address.

Familiarise yourself with all the types of question that you might be asked.

## Worked examples

Detailed worked examples with examiner comments show you how you can raise your grade. Here you will find tips on how to use accurate scientific vocabulary, avoid common exam errors, improve your Quality of Written Communication (QWC), and more.

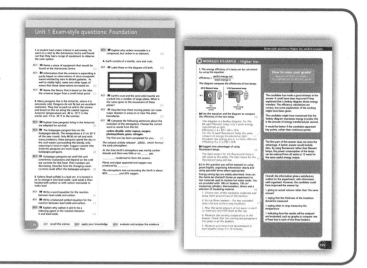

## Preparing for assessment

Each Unit contains Preparing for assessment activities. These will help build the essential skills that you will need to succeed in your practical investigations and Controlled Assessment, and tackle the Assessment Objectives.

Each type of Preparing for assessment activity builds different skills.

> Applying your knowledge: Look at a familiar scientific concept in a new context.

> Planning an investigation: Plan an investigation using handy tips to guide you along the way.

> Analysing and evaluating data: Process data and draw conclusions from evidence. Use the hints to help you to achieve top marks.

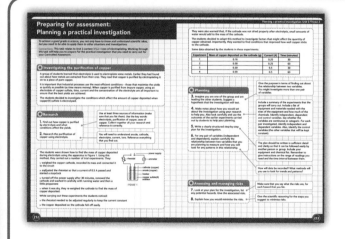

## Bad Science

Based on *Bad Science* by Ben Goldacre, these activities give you the chance to be a 'science detective' and evaluate the scientific claims that you hear everyday in the media.

## Assessment skills

A dedicated section at the end of the book will guide you through your practical and written exams with advice on: the language used in exam papers; how best to approach a written exam; how to plan, carry out and evaluate an investigation; how to use maths to evaluate data, and much more.

# Unit 1 Theme 1: My wider world

## What you should know

### The solar system

We live on a planet called Earth, which is part of the solar system.
White light can be dispersed to produce a spectrum of colours.

● What are the colours of the spectrum of white light?

### The rock cycle

Rocks may be igneous, sedimentary or metamorphic.

Rocks are continually being broken down and built up again in the rock cycle.

● How can an igneous rock become a metamorphic rock?

### Elements and compounds

Elements are made from a single type of atom.

Compounds are made from the atoms of two or more elements.

Atoms are made up of positively charged protons, neutral neutrons (in the nucleus) and negatively charged electrons (outside the nucleus).

● Name some compounds and the elements from which they are made.

### Symbols and formulae

Each atom has a symbol and each compound has a formula.

● Write down the formula for one compound that you have learned about.

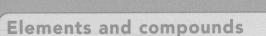

# You will find out

## Our changing universe

> The solar system is part of a galaxy called the Milky Way, which is part of the universe.

> Telescopes that detect electromagnetic radiation can be used to observe the solar system and galaxies.

> Red-shift is evidence of galaxies moving away from each other in an expanding universe.

## Our changing planet

> Earth has a core, mantle and crust.

> The crust and upper mantle form tectonic plates, which move and cause earthquakes when they collide.

> Earth's atmosphere has been changed by volcanoes and photosynthesis to give its present day composition.

> Carbon dioxide is essential to maintain life.

## Materials our planet provides

> Atoms share or transfer electrons to form molecules or ions with a stable electronic configuration.

> Useful materials can be removed from the ground by quarrying or mining.

> Some substances are used straight from the ground but others need to be separated out using processes such as filtration, evaporation and distillation, reduction and electrolysis.

> Fractional distillation separates the gases in air.

## Using materials from our planet to make products

> Chemical symbols and formulae are used to write equations.

> How much material to use, safety, economics and environmental impact must be considered when making a product.

# Earth and beyond

**THE SCIENCE IN CONTEXT** Astronomers, physicists and other scientists have shown that Earth is one of several planets that move around the Sun. This is our solar system. It is part of the Milky Way galaxy, located somewhere within the universe. How did the universe begin? What is it like? How is it changing?

## Robot geologists

NASA launched two robot geologists in 2003. They landed on Mars six months later. Will humans ever land there?

> Earth is part of the solar system, which is part of a galaxy called the Milky Way

> all are part of the universe, which is everything that exists anywhere

### Essential notes

**FIGURE 1**: Mars Rover – one of NASA's robot geologists.

## Our solar system

Look into the night sky – you see thousands of stars. The Sun is a star. Stars are made of gases. They **emit** huge amounts of **energy**. This energy comes from nuclear reactions inside the star.

All planets orbit a star. Earth is a **planet** that moves around a star, the Sun.

Moons are natural satellites that orbit around a planet. Earth has one moon. While Earth orbits the Sun, Earth's moon orbits the Earth.

Exploring space and learning about other planets opens up the possibilities of:

> space travel

> finding useful resources on other planets

> living on other planets.

### Galaxies

A galaxy is lots of stars held together by each others' gravity. Our solar system is part of the Milky Way galaxy and the Sun is one of its 200 billion stars. Galaxies range from millions of stars (dwarf galaxies) to trillions of stars (giant galaxies).

### The universe

The universe is everything that exists anywhere. It is unimaginably big. It is estimated that there are 170 billion galaxies in the universe. Earth is huge compared with some of the tiny creatures that live on it, but it's tiny in the grand scheme of things.

**FIGURE 2**: Our moon is the only place in space where humans have put their feet on the ground. Where else in space have humans 'walked'?

**FIGURE 3**: The Milky Way. It is a dwarf galaxy.

### QUESTIONS

**1** What do all planets orbit?

**2** On how many planets have spacecraft landed?

**3** How many stars are there in the Milky Way?

# Planets in our solar system

Temperatures on planets in our solar system depend on how far they are from the Sun. The further a planet is from the Sun, the colder its surface.

| Planet | Mean distance from the Sun (km) | Mean surface temperature (°C) |
|---|---|---|
| Mercury | 57 900 000 | 167 (varies from -170 to 430) |
| Venus | 108 200 000 | 464 |
| Earth | 149 600 000 | 15 |
| Mars | 227 900 000 | -65 |
| Jupiter | 778 600 000 | -110 |
| Saturn | 1 433 500 000 | -140 |
| Uranus | 2 872 500 000 | -195 |
| Neptune | 4 495 100 000 | -200 |

Venus is the planet nearest to us. It is a similar size to Earth, but extremely hot and dry – and always surrounded by sulfuric acid clouds. Plants and animals that exist on Earth could not live on Venus.

Mercury is unusual. No other planet has the same extremes of temperature. Although very close to the Sun, it spins very slowly. This makes Mercury's day and night much longer than ours. One rotation of Mercury is about 59 Earth days.

FIGURE 4: Venus is sometimes called Earth's 'twin'. Yet, life as we know it cannot exist there. Why not?

## QUESTIONS

**4** How far apart are Mars and Earth?

**5** What is the difference between the highest and lowest temperatures on Mercury?

**6** What is the temperature difference between the surfaces of Earth and Venus?

# Gravity

Gravity is what keeps galaxies together and keeps our feet on the ground. Anything with mass has gravity. It is a force that pulls two objects towards one another. The strength of this pull depends on:

> the mass of the objects: the greater the mass, the bigger the pull

> the distance between the objects: the nearer the two objects, the bigger the pull.

Gravity keeps the planets in orbit around the Sun and moons in orbit around the planets. If the planets were not moving, they would be pulled back into the Sun by its gravity.

### Did you know?

Spacecraft have landed on three planets, two moons (including, of course, our own), two asteroids and one comet.

Q ... gcse gravity

# Exploring space

## Essential notes

> our solar system and galaxies in the universe can be observed from Earth or space

> telescopes that detect electromagnetic radiation such as light, radio waves or X-rays are used

**THE SCIENCE IN CONTEXT**

Many techniques are used to make observations and measurements about our planet, solar system and space beyond. In particular, different types of telescopes, on Earth and in space, provide key data. Scientists analyse and interpret the data to describe space – from neighbouring planets and moons to outer space.

### Galileo

Galileo discovered craters on the moon, sunspots in the Sun and the moons of Jupiter.

**FIGURE 1**: Galileo did not invent the telescope, but he did build the first good ones.

## Looking into space

Without telescopes no one would know much about space. A telescope detects **electromagnetic radiation** from distant objects in space.

Visible light is one form of electromagnetic radiation. A glass prism splits white light to show the spectrum of visible light.

Most electromagnetic radiation is invisible. You feel warmth from infrared radiation and might get sun-burned from **ultraviolet radiation**. Radio waves are used to send radio and TV transmissions.

The whole range of electromagnetic radiation makes up the **electromagnetic spectrum**.

### Optical telescopes

> Optical telescopes detect visible light. Optical telescopes can be used only at night. They cannot be used in cloudy or bad weather.

### Radio telescopes

> Radio telescopes detect the **radio wave** radiation. They can produce images of most objects found in space.

They can be used in bad weather because radio waves are not blocked by clouds. They can be used during the day as well as at night. However, they are very large and expensive.

**FIGURE 2**: Why was the Keck Observatory, with an optical telescope, built high up on Hawaii?

### Remember

An optical telescope detects visible light. A radio telescope detects radio waves.

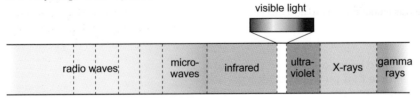

visible light

| radio waves | micro-waves | infrared | ultra-violet | X-rays | gamma rays |

**FIGURE 3**: The electromagnetic spectrum. Apart from visible light, the other types of radiation are invisible to the human eye.

## QUESTIONS

**1** What part of the electromagentic spectrum is detected by an optical telescope?

**2** When can an optical telescope <u>not</u> be used, even in fine weather?

**3** Why can a radio telescope be used even on cloudy days?

# Telescopes and electromagnetic radiation

## Space telescopes

Stars really do twinkle. When light passes through the **atmosphere** it bends because the density of the atmosphere changes. The density decreases the further it is from Earth's surface. It causes the brightness of a star to vary rapidly – this is **scintillation**.

This is why so many optical telescopes are built on mountains, where the atmosphere is less dense. Sharper images are obtained. Problems due to bad weather and clouds are also reduced.

It is difficult and very expensive, but optical telescopes can be put into space. This overcomes the weather and cloud problems. Space telescopes can work day and night, and look in every direction in space. The Hubble space telescope has enabled astronomers to discover new galaxies and observe new stars forming.

## Types of electromagnetic radiation

Telescopes detect electromagnetic radiation. Electromagnetic radiation travels as waves.

Infrared and **X-rays** are blocked by Earth's atmosphere. So space telescopes have to be used to detect them. Two examples are:

> the Chandra X-ray telescope, launched in 1999

> the Spitzer Space infrared telescope, launched in 2003.

**Remember**
Electromagnetic radiation travels as waves and transfers energy from one place to another.

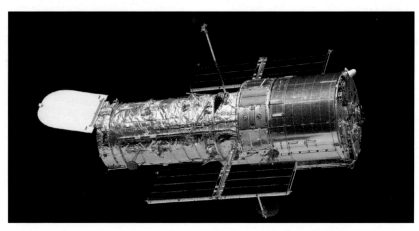

**FIGURE 4**: The Hubble space telescope was launched in 1990 and orbits 59 kilometres above Earth.

## QUESTIONS

**4** Why are sharper images obtained from an optical telescope at the top of a mountain?

**5** What are the disadvantages of space telescopes?

**6** X-ray radiation cannot be detected by instruments on the ground. Why not?

# Big telescopes

The world's largest optical telescope – the European Extremely Large Telescope (E-ELT) – will be 42 metres in diameter when finished. It is being built on a mountain in Chile at an altitude of 3060 metres. There, the skies are clear about 320 nights a year.

The largest radio telescope is the RATAN-600 in Russia. It is almost 600 metres in diameter. Radio waves are not affected by clouds or bad weather, so RATAN-600 can work at an altitude of only 970 metres.

## QUESTIONS

**7** Explain the differences in size and location of E-ELT and RATAN-600.

Q … gcse telescopes

# The expanding universe

**Essential notes**
> the red-shift is evidence for an expanding universe
> galaxies are moving away from one another
> the further away from us, the faster galaxies are moving

**THE SCIENCE IN CONTEXT** Using telescopes to detect light and other electromagnetic radiation from distant galaxies, scientists have developed theories about the nature of the universe – how it began (the Big Bang theory) and how it continues to expand. There is still much to learn!

## Galaxy types

Most galaxies are elliptical, spiral or barred spiral. The others are 'irregular' galaxies.

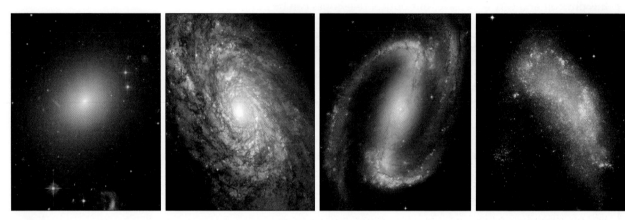

FIGURE 1: From left to right: elliptical, spiral, barred and irregular galaxies.

## Red-shift

### The Doppler effect

As an ambulance comes towards you, its siren makes a high-pitched sound. As it passes and goes away, the sound is low pitched. The sound waves are squashed together as the ambulance comes closer and have shorter **wavelength** and higher pitch. As it goes away the waves are pulled apart and have longer wavelength. This is the **Doppler effect**. The sound that you hear is due to the air vibrating. How fast it vibrates is its frequency.

### Red-shift and the Big Bang

Light waves show a similar effect. Astronomers noticed that light from all galaxies is shifted towards the red end of the visible spectrum. This is called the **red-shift** and is evidence that galaxies are moving away from one another. Hubble discovered that the further a galaxy is away from us, the greater the red-shift. This means the further away a galaxy is the faster it is moving away from us.

George Gamow, a Russian scientist, suggested that if galaxies are moving away from one another, then perhaps they are all travelling from the same starting point. His Big Bang theory describes how this was the starting point of the universe.

FIGURE 2: The number of cycles of a wave that pass in one second is the wave's frequency.

cycle

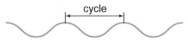

lower frequency

wavelength

higher frequency

wavelength

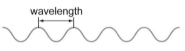

FIGURE 3: The shorter the wavelength of a wave, the higher its frequency.

## QUESTIONS

**1** What did astronomers notice when they observed distant galaxies?

**2** Do all galaxies move at the same speed? Explain your answer.

# Absorption spectrums

### Light from stars and galaxies

Sunlight through a prism shows the visible spectrum as a continuous rainbow of colour. In 1802, William Wollaston noticed dark lines in the spectrum. These are due to elements in the Sun absorbing some of the visible light.

Each **element** has its own characteristic absorption spectrum. The spectrum of any star is a combination of its elements' spectra. Scientists use this to work out the elements in the star.

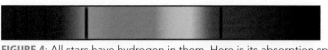

FIGURE 4: All stars have hydrogen in them. Here is its absorption spectrum. (a) How many absorption lines can you see? (b) What colour light is absorbed by hydrogen to give the black line on the right-hand side? Note: there are more lines in the hydrogen spectrum, but they can only be seen using a more sophisticated instrument.

### Evidence for an expanding universe

When a wave source is moving away from an observer, the wavelength appears to stretch and get longer. When moving towards the observer, the wavelength appears to shorten (the Doppler effect). So, when a light source moves away from you it becomes redder – a red-shift. Moving towards you it becomes bluer – a blue-shift.

Hubble looked at the hydrogen spectrum from different stars and galaxies. He noticed that it was the same pattern, but shifted towards the red end of the spectrum. He also observed that the further away the galaxy, the greater the shift into the red.

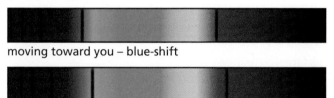

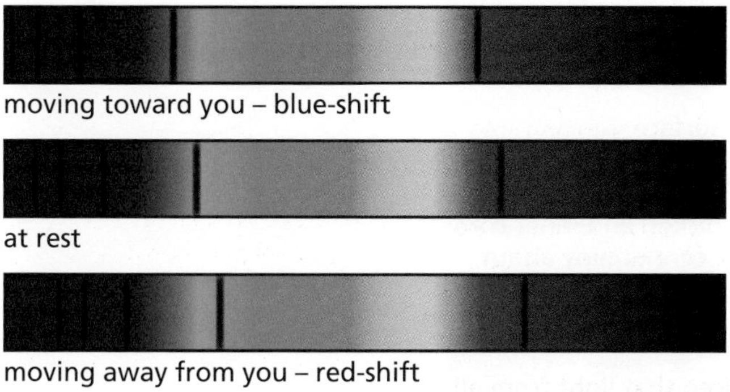

moving toward you – blue-shift

at rest

moving away from you – red-shift

FIGURE 5: Red-shift and blue-shift.

> **QUESTIONS**
>
> **3** In your own words explain red-shift.
>
> **4** Describe two major conclusions that Hubble was able to draw from his work on red-shift.

# Birth of the universe

Scientists believe there was no matter at first – just a gigantic amount of energy concentrated into the size of a pinhead. We do not know where it came from or how it got there. Suddenly, about 14 billion years ago, there was a violent explosion, the Big Bang. The temperature was billions of degrees Celsius, but as the universe expanded it dropped dramatically. Today, its average temperature is -270 °C.

As the temperature decreased, energy changed into matter. After a second (as long as it takes to say 'elephant') basic particles such as quarks, photons and electrons were created. In seconds, protons and neutrons formed. Another 100 000 years later atoms formed.

> **QUESTIONS**
>
> **5** Describe how the atoms that make up the world were created.

# Planet Earth

**THE SCIENCE IN CONTEXT** Earth scientists specialise in studying the Earth's crust. They have gathered evidence that shows Earth has changed a great deal since its formation – and it is still changing. After a period of intense volcanic activity, Earth cooled. It is now able to sustain plant and animal life.

## The Grand Canyon

Earth's surface is constantly changing because of moving tectonic plates and erosion. The Grand Canyon is caused by erosion. It is more than 400 km long, an average 1.6 km deep and between 180 m and 28.8 km wide.

FIGURE 1: The Grand Canyon.

## Planet Earth

Earth is about 4.6 billion years old. Its surface is made of rock. Just over 70% is covered in water (seas, oceans and rivers). Around Earth is a very thin layer of air (Earth's atmosphere).

On Earth there are mountains, deserts, rainforests, polar ice caps, islands and much more. It was not like this when Earth first formed. Earth has changed and continues to change, but too slowly to notice (apart from the occasional volcanic eruption or **earthquake**).

### The surface and below

Earth has three layers. From the outside to the middle they are:

> crust – solid rock and relatively thin; this is Earth's surface

> mantle – molten rock (called magna) and about half Earth's diameter

> core – consisting of an outer core (liquid iron and nickel) and inner core (solid iron and nickel); together the two parts of the core occupy about half Earth's diameter.

**Remember**
Earth is very nearly a sphere. Its surface is rocky and 70% is covered with water. Around it is the atmosphere.

FIGURE 2: Earth is sometimes called the blue planet because water on its surface looks blue when seen from space.

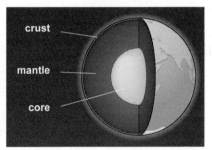

crust

mantle

core

FIGURE 3: The layers that make up Earth.

## QUESTIONS

**1** What percentage of Earth's surface is covered by water?

**2** Name the three main layers of Earth and say which two layers are thickest.

**3** Name the two parts of Earth's core and describe how they differ.

# Crust, mantle and core

## The crust and mantle

Take away the water and you would see that a rocky crust covers the planet. However, the crust is broken into large pieces called **tectonic plates**. The upper part of the mantle forms part of these tectonic plates.

The plates are constantly moving, about a centimetre or two each year (about the same speed as your fingernails grow). Over millions of years, whole continents have shifted thousands of kilometres apart. This is **continental drift**.

## The core

The core is believed to be mainly iron with about 4% nickel. At the very centre of Earth the mixture is solid (the inner core). This is surrounded by a liquid outer core.

Earth is magnetic. Scientists believe the magnetic field is generated in the outer core.

Nobody has ever been to the core. No one has drilled much more than twelve kilometres below the surface. Evidence about the composition of the core comes from:

> the shape of Earth's surface, including the crust beneath the oceans

> rock that has been brought from deep down to the surface by volcanic eruptions

> seismology – analysis of how sound waves behave when they travel through Earth

> gravity measurements.

**Remember**

Earth's crust is not continuous. It is made up of large, slow-moving tectonic plates.

FIGURE 4: You can use a compass to find your way. Its metal needle lines itself up with Earth's magnetic field. Which part of Earth generates its magnetic field?

## QUESTIONS

**4** How far have tectonic plates have moved in your lifetime?

**5** Describe the difference between Earth's inner core and outer core.

# Earth loses its magnetism

Over Earth's life, the magnetic field strength has steadily decreased, disappeared and switched over.

At the moment, it is fading again. It has lost 10% of its strength in the last 150 years and may disappear in 1500 to 2000 years. It will gradually come back, but today's compasses will point south instead of north.

While there is no magnetic field, Earth is vulnerable to the electrically charged particles that stream from the Sun in solar winds. These can damage living cells and could cause the extinction of many organisms.

## QUESTIONS

**6** Explain what solar wind is and how Earth protects us from it.

Q ... gcse the earth's magnetic field

# Tectonic plates

**THE SCIENCE IN CONTEXT** From their investigations, scientists have evidence that the ground you stand on is constantly moving – though only very slowly. Pieces of Earth's crust, called tectonic plates, move around. Colliding plates often cause earthquakes – some small and some huge. Geophysicists and seismologists study earthquakes, learning about their causes and trying to predict when they might happen.

### Killer waves

Earthquakes can cause huge waves to form – called tsunamis. In 2004, a devastating tsunami killed more than 240 000 people.

**FIGURE 1**: Earthquakes can cause tsunamis, often with devastating effects.

## Moving plates

Earth's surface is made up of **tectonic plates**. They are packed together. Importantly, they are always moving, though very slowly.

Tectonic plates move because of **convection currents** in Earth's mantle. They 'ride' on top of Earth's mantle.

In a room, warm air rises, cools and sinks – as it does, warm air from below rises and takes its place. This causes **convection** currents that circulate air around the room. In Earth's mantle, radioactive changes release energy that heats the molten rock. This starts convection currents.

Moving plates collide. As they push against one another, the force gradually increases until, suddenly, one plate pushes up and rides over the other. There is an enormous release of energy and an earthquake happens.

streaks of purple dye moving through clear water

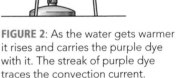

**FIGURE 2**: As the water gets warmer it rises and carries the purple dye with it. The streak of purple dye traces the convection current.

### QUESTIONS

1 Name the large pieces that form Earth's crust.

2 What causes convection currents in Earth's mantle?

3 Explain briefly how an earthquake happens.

## Plates on the move

Tectonic plates may be sliding:

> towards one another (a convergent boundary) – collision will cause an earthquake and Earth's crust may scrunch up and form mountain ranges

> away from one another (a divergent boundary) – molten rock (magna) from the mantle pushes up through the gap and forms new crust

> past one another (a transform boundary) –  may also cause an earthquake.

Q ... tectonic plates

FIGURE 3: High in the Himalayan mountains there are fossilised sea creatures. At one time, India was an island in the sea. It collided with part of Asia, which included Tibet, and created the Himalayas. Explain how the mountains formed and how the fossils came to be high in the mountains.

**Remember**
Earthquakes happen when tectonic plates collide and, sometimes, when they grind past one another.

Two plates sliding or grinding past one another (a transform boundary) may cause an earthquake. The San Andreas Fault in California is an example. The last notable earthquake there happened in 2004. It had a magnitude of 6.0 and was due to a sudden plate movement of about 50 centimetres.

## QUESTIONS

**4** What is the difference between a convergent boundary and a transform boundary?

**5** Explain how earthquakes at the San Andreas Fault might happen.

## Continents and land masses

In 1915, Alfred Wegener suggested that a supercontinent, **Pangaea**, existed in the past and broke up 200 million years ago. He proposed that the pieces 'drifted' to their present positions. At the time, most scientists thought that he was wrong.

By the 1960s, there was clear evidence to support his theory:

> Earthquakes and volcanoes happen in specific places – at the boundaries of moving plates.

> The east coast of South America and the west coast of Africa fit together neatly – suggesting they were once joined.

> In places, rock patterns and fossils (including land animals that could not swim) are the same on both sides of the Atlantic.

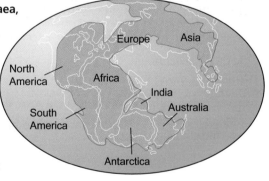

FIGURE 4: Pangaea together with outlines of the continents and land masses that exist now. Notice how they fit together like a huge jigsaw puzzle.

## QUESTIONS

**6** Describe how the continents and land masses that exist now on Earth might have formed from the supercontinent, Pangaea.

# Volcanoes

**Essential notes**

- > erupting volcanoes helped create Earth's atmosphere
- > volcanoes are still erupting, usually linked to tectonic plates colliding
- > volcanic eruptions may be predicted, but not with great accuracy

**THE SCIENCE IN CONTEXT** Earth's surface continues to change because of earthquakes and volcanic eruptions. These happen mainly along the edges of tectonic plate boundaries. Volcanologists study volcanoes. Inactive volcanoes are studied by geologists. Being able to predict earthquakes and volcanic eruptions could help to reduce the devastation that they cause.

### Volcanic ash

When the Icelandic volcano, Eyjafjallajökull, erupted in 2010, it produced a nine kilometre high ash cloud.

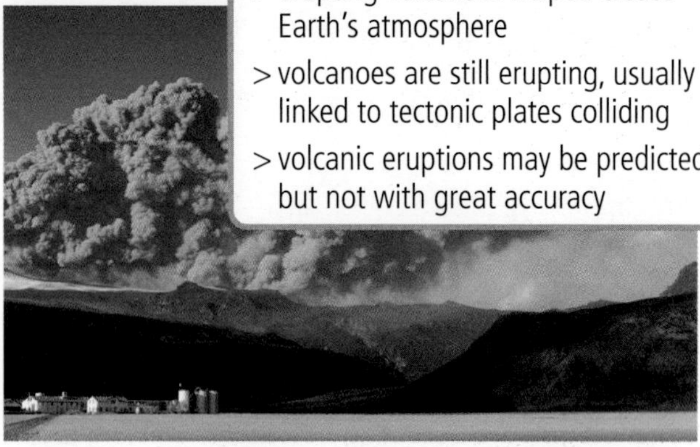

**FIGURE 1**: Iceland spans the North American Plate and the Eurasian Plate.

## Volcanoes old and new

### Early volcanoes

About 4.5 billion years ago, when Earth was a ball of **molten** rock, it collided with something.

Earth then cooled rapidly. Its first atmosphere (possibly hydrogen and helium) formed, but disappeared quickly into space.

As Earth cooled, a crust formed. During Earth's first billion years, violent volcanic eruptions released molten rock and gases (mainly steam, **carbon dioxide**, **ammonia** and **methane**). These gases formed Earth's second atmosphere.

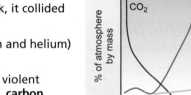

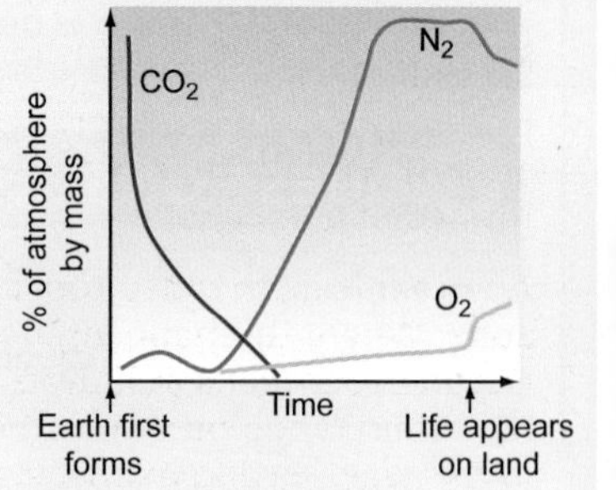

**FIGURE 2**: Changes in Earth's atmosphere.

With further cooling, steam condensed to make water. Seas and oceans formed. Carbon dioxide dissolved in these waters and so its concentration in the atmosphere decreased.

Over time, ammonia from volcanoes formed nitrogen. Later, plants began to grow and produce oxygen. The proportions of nitrogen and oxygen now in the air resulted from a balance between the two processes.

### Today's volcanoes

**Volcanoes** most often happen at plate boundaries. They form when bubbles of hot molten rock (**magma**), work their way up from the mantle and force their way through Earth's crust.

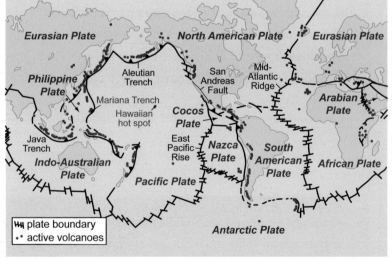

**FIGURE 3**: The main plate boundaries and active volcanoes.

### QUESTIONS

**1** Which gases, produced by volcanic eruptions, formed Earth's 'second' atmosphere?

**2** Describe the relationship between plate boundaries and volcanoes.

# Predicting eruptions

The tell-tale signs of a volcanic eruption:

> Molten rock moves beneath the volcano. This creates small earthquakes or tremors. The ground 'shakes' and the vibrations are detected by a seismometer.

> The molten rock collects in a magma chamber. It causes the slopes of the volcano to bulge very slightly. This is measured using tiltmeters and geodimeters (measuring the distance between two points).

> As it gets closer to the surface, the magma releases gases such as **sulfur dioxide**. Spectrometers are used to identify the gases and measure their concentrations.

Volcanologists are becoming better at predicting eruptions. However, there are still difficulties.

> It is very hard to say exactly when an eruption will happen. For example, moving magma often cools beneath the surface and there is no eruption.

> Monitoring is expensive.

> It is impossible to monitor every site.

## Hot spots and volcanic eruptions

Hot spots are where magma pushes tectonic plates apart at their boundaries and comes to the surface. The Hawaiian Islands formed at a **hot spot**.

An **oceanic plate** slides under a **continental plate** at a **subduction zone**. Where this happens, magma can rise to form a volcano. Oceanic rock melts as it dives into the mantle. The Andes mountain range on the west coast of South America was formed in this way, leaving extinct volcanoes dotted along the coast.

**Remember**
Scientists can predict volcanic eruptions with increasing accuracy, but there are still uncertainties.

### Did you know?

Millions of people are threatened by potential volcanic eruptions. Being able to predict eruptions is really important.

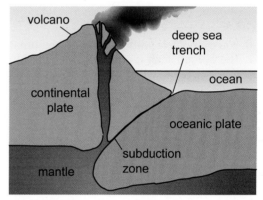

**FIGURE 4**: Forming a subduction zone. An oceanic plate is pushed down and under a continental plate when they collide.

## QUESTIONS

**3** Describe three signs that volcanologists look for when checking for possible eruptions.

**4** Why doesn't a small earthquake always signal a volcanic eruption?

**5** Why does an oceanic plate move under a continental plate?

# The deepest place in the oceans

The deepest place in the oceans is the Mariana Trench. It is just east of the Philippines in the Pacific Ocean. How did it form?

The Mariana Trench marks where the fast-moving Pacific plate pushes against the slower-moving Philippine Plate. The heavier oceanic plate is sliding downward towards the molten mantle. The lighter, continental plate rides up over the top. The really deep part of the ocean is at the bottom of the trench created by the subducting oceanic plate.

## QUESTIONS

**6** Why are volcanoes often found near deep sea trenches?

# Oxygen

**THE SCIENCE IN CONTEXT** Earth's atmosphere is unique in the solar system. No other planet has oxygen in its atmosphere, and oxygen is essential to life as we know it. Plant biologists and botanists study plant propagation and growth. They have worked out how plants use sunlight when they make organic compounds and oxygen. The process is called photosynthesis.

## Oxygen the killer

2.5 billion years ago there was little or no oxygen in the atmosphere, yet there was primitive life. When oxygen began to accumulate in the atmosphere, it was like poison to these primitive organisms – vast numbers of species became extinct.

**FIGURE 1**: Cyanobacteria produce oxygen by photosynthesis.

## Photosynthesis

You cannot live without oxygen. You breathe in air. The oxygen in air is absorbed into your blood. It is then carried to wherever there is work to be done.

Oxygen is produced by algae and plants, by **photosynthesis**.

carbon dioxide + water + <u>energy</u> ➡ glucose + oxygen

The energy for photosynthesis comes from the Sun as light – electromagnetic radiation in the visible region of the electromagnetic spectrum.

About 2.5 billion years ago, some types of **bacteria** started to release significant quantities of oxygen into the atmosphere. They did this by photosynthesis and are called photosynthetic bacteria.

As living organisms evolved, more algae and plants began to exist, producing more oxygen.

**FIGURE 2**: Green leaves absorb sunlight.

### Green leaves

Green leaves contain chlorophyll. This is the chemical **compound** that absorbs sunlight. Sunlight provides the energy for photosynthesis.

### QUESTIONS

**1** What is the process by which plants make glucose from carbon dioxide and water?

**2** When were significant quantities of oxygen gas first released into the atmosphere?

**3** What is the compound in green leaves that absorbs sunlight?

## Respiration

You need oxygen to produce glucose. The **chemical reaction** between glucose and oxygen releases energy in the process of **respiration** (or, more accurately, aerobic respiration):

glucose + oxygen ➡ carbon dioxide + water + <u>energy</u>

The reaction is the reverse of photosynthesis.

However, Earth's atmosphere has not always contained oxygen gas.

### Did you know?

Earth is the only planet in our solar system that has oxygen in its atmosphere.

## Before there was oxygen in the atmosphere

When oxygen gas started to be released in photosynthesis, it reacted with iron compounds. Evidence for this are the layers of iron oxide, called banded iron formations, which have been found in sedimentary rocks. When there was little or no oxygen in the atmosphere, primitive organisms survived by anaerobic respiration:

glucose ➡ **ethanol** + <u>energy</u>

About 2.5 billion years ago, the Great Oxygenation Event happened. Before the GOE, dissolved iron compounds reacted with oxygen dissolved in the oceans. As the iron compounds became used up, the oxygen started to escape into the atmosphere.

Oxygen was like poison to the huge numbers of primitive organisms and they became extinct. Organisms that used aerobic respiration took over. Eventually this led to the **evolution** of plants and animals as they are today.

**Remember**
Aerobic means *depend* on oxygen. Anaerobic means *does not depend* on oxygen.

**Remember**
Aerobic organisms need oxygen to survive. Anaerobic organisms do not need oxygen.

FIGURE 3: How were these banded iron formations made?

FIGURE 5: What eventually led to the evolution of these plants and animals?

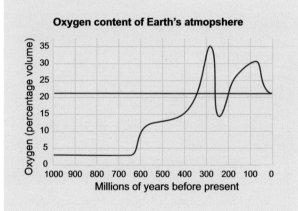

FIGURE 4: What was the percentage of oxygen in Earth's atmosphere (a) 800 million years ago (b) 500 million years ago (c) 250 million years ago (d) 100 million years ago?

### QUESTIONS

**4** What is the difference between anaerobic organisms and aerobic organisms?

**5** When were oxygen levels in Earth's atmosphere at their highest?

## The rise and fall of oxygen

Oxygen levels began to rise about 600 million years ago, coinciding with the appearance of multicelluar animals. Oxygen concentrations were at their highest 300 million years ago, during the carboniferous period. This was the time of vast swamps in which most of the fossil fuels used today were formed.

Oxygen concentrations fell around 250 million years ago, as a result of Siberian volcanoes erupting. For 5 million years they spewed acidic ash into the atmosphere. It created acid rain that killed off much of life on Earth.

FIGURE 6: An artist's impression of the carboniferous period.

### QUESTIONS

**6** Suggest reasons why concentrations of oxygen:
(a) increased during the carboniferous period
(b) decreased during the five million years of volcanic eruptions. Hint: look at Figure 4.

# A changing atmosphere

THE SCIENCE IN CONTEXT Earth's early atmosphere had lots of carbon dioxide, but no oxygen. Over time this changed – carbon dioxide concentrations decreased and oxygen increased. Nowadays, carbon dioxide is only 0.03% of the atmosphere, but it has a vital role in keeping Earth's surface warm. Environmental scientists are beginning to understand this role and its effect on climate change.

## Sinking the Titanic

Icebergs can be spectacular, but dangerous! The Titanic was famously sunk by colliding with an iceberg.

FIGURE 1: Melting ice sheets could lead to rising sea levels.

### Essential notes

> Earth's atmosphere changed over time before reaching its present composition

> carbon dioxide in the atmosphere is essential in sustaining life because it traps heat from the Sun, helping keep Earth's surface warm.

## Changes over time

Following volcanic eruptions, Earth's 'second' atmosphere consisted of water vapour, carbon dioxide, ammonia and methane. Over time:

> Water vapour condensed to form Earth's earliest oceans.

> Carbon dioxide dissolved in the oceans. It became 'locked up' in sedimentary rocks such as **limestone**.

> Oxygen gas began to form when carbon dioxide was photosynthesised. At first this was by certain types of bacteria and later by green plants.

> Nitrogen gas was produced partly by ammonia from volcanoes reacting with oxygen and partly by organisms such as denitrifying bacteria.

This led to the atmosphere that we have now – mainly nitrogen and oxygen. Often, it is simply called air. Its composition, based on 'dry' air, is shown in the table.

| Gas | Formula | % of the atmosphere |
|---|---|---|
| nitrogen | $N_2$ | 78.1 |
| oxygen | $O_2$ | 20.9 |
| argon | Ar | 0.9 |
| carbon dioxide | $CO_2$ | 0.03 |
| others | | 0.07 |

## QUESTIONS

**1** What gases were produced by Earth's early volcanic eruptions?

**2** How was carbon dioxide removed from the atmosphere?

**3** How did the proportion of oxygen in the atmosphere go up?

## Carbon dioxide

### Carbon dioxide in the atmosphere

Carbon dioxide is produced by animals and plants in aerobic respiration. This is a series of chemical reactions in cells that can be summarised as:

glucose + oxygen ➡ carbon dioxide + water + energy

Other natural processes such as forest fires and volcanic eruptions also put carbon dioxide into the atmosphere. Since the Industrial Revolution, burning fossil fuels has added carbon dioxide faster than it can be removed.

**fossil fuel** + oxygen ➡ carbon dioxide + water + energy

### Remember

Earth's atmosphere is often called 'air'. Many planets have atmospheres, but their compositions are different and so we do not call them air.

## The importance of carbon dioxide

The proportion of carbon dioxide in our atmosphere is very small, but it has a vital role in sustaining life. It traps heat from the Sun and helps keep Earth's surface warm. What happens:

> Radiation from the Sun passes through Earth's atmosphere.

> Heat is emitted back from Earth's surface.

> Some of this heat passes through the atmosphere and back into space.

> Some is absorbed by carbon dioxide and keeps Earth's surface warm enough to support life.

Ideally, the amount of greenhouse gas in the atmosphere should be enough to just balance what is needed to keep the temperature steady.

The more heat that is trapped in this way, the warmer the air becomes and, therefore, Earth's surface as well. Other gases that do this are called **greenhouse gases**. Water vapour and methane are two examples.

There is strong evidence that the average global air temperature is rising. This could lead to **melting** ice, rising sea levels and different weather patterns (hotter summers, colder winters and more storms).

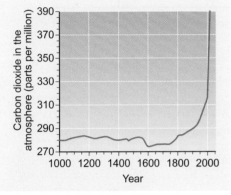

FIGURE 2: The Industrial Revolution began early in the 18th century.

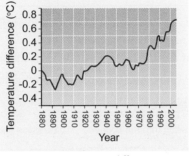

FIGURE 3: Temperature difference compared with the 1880 average. How much did the average temperature of Earth's atmosphere increase between 1900 and 2000?

### QUESTIONS

**4** Name two natural processes that put carbon dioxide into the atmosphere.

**5** Why did the Industrial Revolution affect the proportion of carbon dioxide in the atmosphere?

**6** Explain the importance of having carbon dioxide in our atmosphere.

# How do greenhouse gases work?

Radiation in the visible region of the electromagnetic spectrum, with small amounts of infrared radiation and ultraviolet radiation, passes through Earth's atmosphere.

Light radiation has short wavelengths and, therefore, high energy. Earth's surface absorbs this energy and emits it again as infrared radiation. This has a longer wavelength and carries less energy.

Some infrared radiation is absorbed by greenhouse gases and some escapes through the atmosphere back into space.

### QUESTIONS

**7** Use scientific language to describe what the diagram in Figure 4 shows.

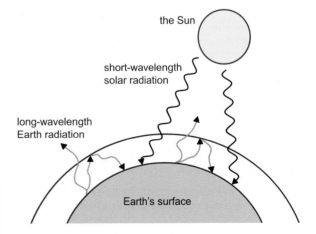

FIGURE 4: The greenhouse effect.

# Preparing for assessment:
# Processing and analysing data

*To achieve a good grade in science, you not only have to know and understand scientific ideas, but you need to be able to apply them to other situations and investigations.*

Connections: This task relates to Unit 1 context 3.3.1.2 Our changing planet. Working through this task will help you to prepare for the practical investigation that you need to carry out for your Controlled Assessment.

## ✳ Investigating the composition of air

A group of students decided to investigate the composition of air. They had been told that the composition of 'dry' air is:

| Gas | nitrogen | oxygen | argon | carbon dioxide | others |
|---|---|---|---|---|---|
| Formula | $N_2$ | $O_2$ | Ar | $CO_2$ | |
| % by volume of the air | 78.1 | 20.9 | 0.9 | 0.03 | 0.07 |

The students decided to find the percentage of oxygen in air. They did this by passing air over heated copper turnings.

Oxygen reacts with copper at high temperatures to form copper oxide: $O_2 + 2Cu \rightarrow 2CuO$

The decrease in volume shows how much oxygen was in the original volume of air.

This is the apparatus they used:

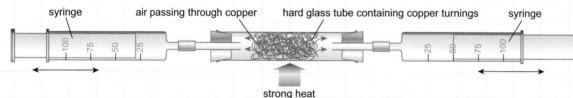

strong heat

Initially, one syringe had some air in it and the other had no air in it. The copper turnings were heated strongly and air was passed to and fro over them by pushing it from one syringe to the other and back again several times. After the last pass, the apparatus was left to cool and the volume of gas remaining was measured.

Their teacher pointed out two things to remember:

> The volume of a gas increases with temperature. So, it is important to let the apparatus cool down at the end of the experiment, otherwise the final reading would be too high.

> The oxygen in the air in the glass tubing containing copper turnings will also react with the hot copper, causing a small error in the final volume recorded.

Each student did the experiment twice. Here are their results:

> In Katie's first experiment, she started with 100 cm³ of air and ended up with 75.2 cm³. Second time, she started with 80.0 cm³ and finished with 64.4 cm³.

> In Indu's first experiment, he started with 40.0 cm³ of air and finished with 29.8 cm³. In his second experiment, 90.0 cm³ reduced to 70.2 cm³ after the air had been passed to and fro.

> In Ellie's first experiment she began with 85.0 cm³ of air and finished with 66.2 cm³. The second time, she started with 55.0 cm³ and finished with 41.6 cm³.

> In Cris's first experiment she started with 100.0 cm³ of air and finished with 81.8 cm³. In her second experiment she found that 78.4 cm³ were left from the 100.0 cm³ that she started with.

## ✸ Processing data

**1.** Construct and complete a results table for the students' experiments.

> Make sure that the headings of the rows and columns are complete, and that you state the units.

**2.** Look at the results table and identify the result that appears to be most obviously anomalous.

> Remember that *anomalous* means it does not fit the pattern shown by the other data.

## ✸ Analysing data

**3.** Using the results obtained from the students' experiments, calculate the volume of oxygen that was used up by the reaction with the hot copper turnings. Draw a stacked bar chart to show the proportion of oxygen to the other gases in air for each experiment.

> You can draw the stacked bar chart by hand or using suitable software such as an Excel spreadsheet. You need to label both axes.
>
> Remember, a stacked bar chart shows two values that make up the total.

**4.** On your stacked bar chart, add a label to say which data are continuous and which are categoric.

> You may need to look up the terms *continuous data* and *categoric data*.

**5.** For each experiment, calculate the percentage by volume of oxygen in air. Then draw a bar chart of the results from the eight experiments and label the anomalous result.

> Take care to clearly state your answers using the appropriate number of significant figures. Does the result agree with what you thought in question 2?

**6.** Ignoring the anomalous result, write down the highest and lowest values for percentage oxygen in air and calculate the mean value. Compare this with the value that you had been given.

> Remember that the group's results are primary data, but the composition of air you were given at the beginning is secondary data.

**7.** Look at your column diagram and identify the two experiments where it is likely that the students did not let the apparatus cool completely before measuring the volume of the remaining gas.

> Apart from the anomalous result, which two other results are higher than the others (to be expected if the gas was still hot)?

**8.** Calculate the mean value for the percentage oxygen in air, omitting the anomalous result and the two where it is likely that the apparatus had not been cooled.

> What difference did it make if you omitted the three results?

# Elements, compounds and mixtures

**THE SCIENCE IN CONTEXT** Useful substances come from Earth's crust, sea and atmosphere. Rocks, minerals and fossil fuels are among the raw materials used to make everything we need. Virtually all are mixtures. They have to be extracted, separated and purified. Large numbers of scientists, technologists and engineers are involved.

### The first modern chemistry textbook

In 1789, Antoine Lavoisier published 'Elementary Treatise of Chemistry'. His list of elements included oxygen, nitrogen, hydrogen, zinc and sulfur. However, it also included light and heat. He believed that these were materials. He did not know that they are ways of transferring energy.

**FIGURE 1**: Earth provides all we need to live, reproduce and survive.

---

## Elements and compounds

### Elements

Look at a gold ring or a diamond and you are looking at elements. Breathe in and your lungs fill with invisible elements.

Most of the world is not like that. Mostly it consists of compounds.

Everything around you is made from tiny particles called atoms. An element has just one type of **atom**. Atoms of one element are different from the atoms of all other elements.

Each element has a **symbol**. For example, hydrogen is H, carbon is C and sodium is Na.

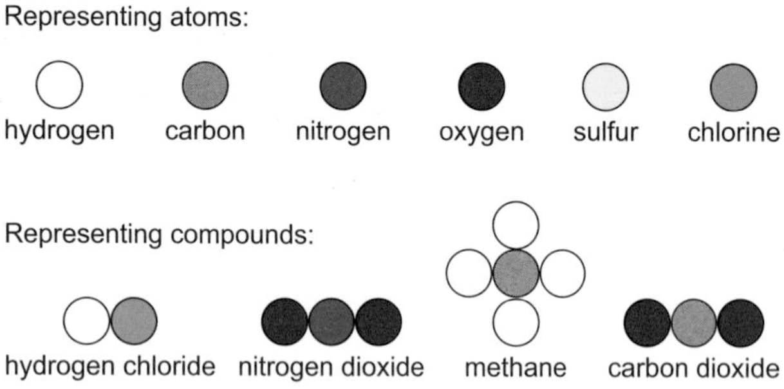

Representing atoms:

hydrogen   carbon   nitrogen   oxygen   sulfur   chlorine

Representing compounds:

hydrogen chloride   nitrogen dioxide   methane   carbon dioxide

**FIGURE 2**: Each circle represents an atom – a different colour for each element. You may find this a useful way to identify and count atoms.

### Did you know?

Only a few elements were known in 1806 – and some of these turned out not to be elements.

### Compounds

Compounds are substances made from atoms of two or more elements joined together. Although there are only about 100 elements, different combinations of their atoms result in countless numbers of compounds.

Elements and compounds are pure chemical substances, but they rarely exist alone. Two or more pure substances are mixed together. Rocks, **crude oil** and air are examples of mixtures. Mixtures can be separated by physical processes such as filtration and **distillation**.

Each compound has a **formula** that shows the types and proportions of atoms from which it was made. For example, the formula for carbon dioxide is $CO_2$. This tells you that carbon dioxide is made from carbon atoms and oxygen atoms in the proportion 1:2.

> **Remember**
> Ninety elements occur naturally. A few more are made artificially.

### QUESTIONS

1 What is special about an element?

2 How many different types of element are there in nature?

3 What is the difference between an element and a compound?

# Mixtures and compounds

## Separating mixtures

A mixture contains two or more substances that are not combined chemically.

Mixtures can be separated into simpler mixtures with fewer components. Imagine a bucket of seawater on a sand beach. Your bucket has a mixture of components:

> water

> various solids such as sodium chloride dissolved in the water

> sand stirred up by waves and currents.

Filter the sand and you are left with a solution. This is still a mixture – water and dissolved solids.

Evaporate the solution and you are left with a solid. This is still a mixture – its components are the various dissolved solids.

Keep on separating and you end up with pure substances (elements or compounds).

FIGURE 3: Sea salt is mainly sodium chloride. It is made by evaporating seawater.

## Decomposing compounds

A compound can be broken down into simpler compounds. For example, if you heat **calcium carbonate**, it decomposes into calcium oxide and carbon dioxide.

calcium carbonate ➡ calcium oxide + carbon dioxide

However, there comes a point at which a product cannot be broken down further. The compound has been broken down into its elements. Passing an electric **current** through a molten compound or through a solution of the compound can be used in some cases. This is called **electrolysis**.

Two examples of compounds that can be broken down by electrolysis:

water ➡ hydrogen and oxygen

lead chloride ➡ lead + chlorine

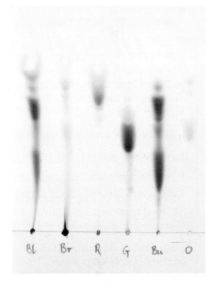

FIGURE 4: A mixture of coloured compounds can be separated using chromatography.

## QUESTIONS

**4** When calcium carbonate is heated it gives off a gas. Does this show that calcium carbonate is an element or a compound?

# Air

Air is a mixture. Some of the substances are elements and some are compounds.

| Gas | Element or compound | Particles | Formula | Each particle made from |
|---|---|---|---|---|
| nitrogen | element | molecules | $N_2$ | two nitrogen atoms |
| oxygen | element | molecules | $O_2$ | two oxygen atoms |
| argon | element | atoms | Ar | one argon atom |
| carbon dioxide | compound | molecules | $CO_2$ | one carbon atom and two oxygen atoms |

There are also water vapour ($H_2O$), trace amounts of methane ($CH_4$) and pollutants such as sulfur dioxide ($SO_2$). In each case the particles are molecules.

## QUESTIONS

**5** Write down what each of these molecules is made from (a) water (b) methane (c) sulfur dioxide.

Q ... gcse mixtures

# Atoms

**THE SCIENCE IN CONTEXT** We can find out about materials empirically – by 'trial and error'. Changes can be observed, patterns of behaviour recognised and predictions made on further changes. An understanding of chemical structure enables scientists to make best use of them. The starting point is the structure of atoms – the building blocks of all materials.

## First periodic table

In 1869 Dimitri Mendeléev, a Russian chemist, produced the first periodic table.

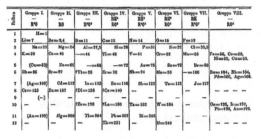

**FIGURE 1**: Mendeléev's periodic table.

## Sub-atomic particles

Line up 30 million carbon atoms and they would only stretch about one centimetre. Atoms are really tiny – but they are not the smallest particles. Atoms are made up of sub-atomic particles – protons, neutrons and electrons.

> A **proton** has a positive **charge** (+1).

> A **neutron** has no charge (0).

> An **electron** has a negative charge (-1).

At the centre of every atom is a **nucleus**. It contains protons and neutrons. The nucleus is surrounded by electrons that move around it.

The number of electrons in an atom is always the same as the number of protons. Their charges cancel out, so an atom has no overall charge.

### Atoms and elements

The atoms of a particular element have the same number of protons as each other. For example, all carbon atoms have six protons, all oxygen atoms have eight protons and all iron atoms have 26 protons.

Atoms of one element have a different number of protons from atoms of every other element.

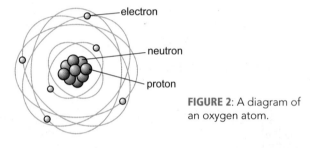

**FIGURE 2**: A diagram of an oxygen atom.

### QUESTIONS

**1** What is at the centre of an atom? Which sub-atomic particles does it consist of?

**2** What type of charge does an electron have?

**3** Explain why an atom has no charge.

### Did you know?

Atoms began to form about 300 000 years after the Big Bang.

## Numbers, tables and arrangements

### Atomic number

The number of protons is called the atomic number. Each element has a unique atomic number.

| Element | hydrogen | carbon | oxygen | sodium | chlorine | iron |
|---|---|---|---|---|---|---|
| **Atomic number** | 1 | 6 | 8 | 11 | 17 | 26 |

**TABLE 1**: Examples of atomic numbers.

## Mass number

The sum of the number of protons and neutrons in an atom is its **mass number**.

An oxygen atom has eight protons and eight neutrons, so its mass number is 8 + 8 = 16.

If you know the atomic number and mass number of an atom, the numbers of its sub-atomic particles can be calculated.

A sodium atom has an atomic number of 11 and a mass number of 23. Therefore it has:

> 11 protons and 11 electrons (from the atomic number)

> (23 − 11) = 12 neutrons (taking the atomic number away from the mass number)

Atoms of the same element that have different mass numbers are called isotopes.

## The periodic table

In the modern **periodic table**, elements are arranged in order of increasing atomic number. Elements with similar properties are arranged in vertical columns called **groups**. The rows are called **periods**. Each period has metals at the left and non-metals at the right, divided by the red zig-zag line.

| Group | | | | | | | | | | | | | | | | | 0 |
|---|---|---|---|---|---|---|---|---|---|---|---|---|---|---|---|---|---|
| 1 | 2 | | | | | | | | | | | 3 | 4 | 5 | 6 | 7 | He |
| | | H | | | | | | | | | Group | | | | | | |
| Li | Be | | | | | | | | | | | B | C | N | O | F | Ne |
| Na | Mg | | | | | | | | | | | Al | Si | P | S | Cl | Ar |
| K | Ca | Sc | Ti | V | Cr | Mn | Fe | Co | Ni | Cu | Zn | Ga | Ge | As | Se | Br | Kr |
| Rb | Sr | Y | Zr | Nb | Mo | Tc | Ru | Rh | Pd | Ag | Cd | In | Sn | Sb | Te | I | Xe |
| Cs | Ba | La | Hf | Ta | W | Re | Os | Ir | Pt | Au | Hg | Tl | Pb | Bi | Po | At | Rn |
| Fr | Ra | Ac | | | | | | | | | | | | | | | |

**FIGURE 3**: The periodic table.

## Electron arrangements

Electrons are positioned outside the nucleus. They are arranged in layers called **shells**. Electrons occupy the lowest available **energy levels**.

> The first shell can hold up to two electrons.

> The second shell can hold up to eight electrons.

> The third shell can hold up to 18 electrons.

The arrangement of electrons in these shells is called the **electronic configuration**.

The **atomic number** of sodium is 11, so each sodium atom has 11 electrons. Two are in the first shell, eight in the second shell and one in the third shell. This is written as 2.8.1 and may be shown using a diagram:

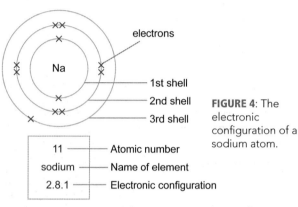

**FIGURE 4**: The electronic configuration of a sodium atom.

## QUESTIONS

**4** The atomic number of nitrogen is 7. It mass number is 14. How many of the following does it contain (a) protons (b) neutrons (c) electrons?

**5** What is the electronic configuration of sulfur (atomic number 16)?

# Electronic configurations and the periodic table

The properties and chemical reactions of elements depend on their electronic configurations. This explains their arrangement in the periodic table – elements in a **group** have the same number of electrons in their outer shell.

## QUESTIONS

**6** Helium, neon and argon do not react with anything. Suggest why.

| 1 | | | | | | | 2 |
|---|---|---|---|---|---|---|---|
| hydrogen | | | | | | | helium |
| 1 | | | | | | | 2 |
| 3 | 4 | 5 | 6 | 7 | 8 | 9 | 10 |
| lithium | beryllium | boron | carbon | nitrogen | oxygen | fluorine | neon |
| 2.1 | 2.2 | 2.3 | 2.4 | 2.5 | 2.6 | 2.7 | 2.8 |
| 11 | 12 | 13 | 14 | 15 | 16 | 17 | 18 |
| sodium | magnesium | aluminium | silicon | phosphorus | sulfur | chlorine | argon |
| 2.8.1 | 2.8.2 | 2.8.3 | 2.8.4 | 2.8.5 | 2.8.6 | 2.8.7 | 2.8.8 |
| 19 | 20 | | | | | | |
| potassium | calcium | | | | | | |
| 2.8.8.2 | 2.8.8.2 | | | | | | |

**FIGURE 5**: Electronic configurations of the first 20 elements.

# Molecules and ions

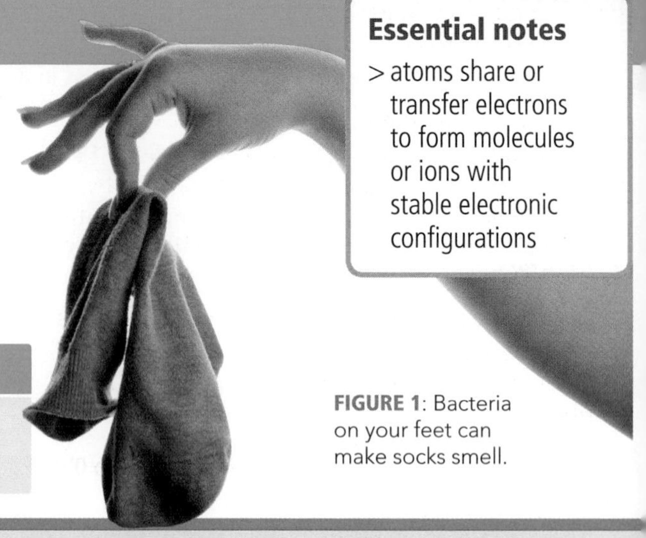

FIGURE 1: Bacteria on your feet can make socks smell.

**Essential notes**
- > atoms share or transfer electrons to form molecules or ions with stable electronic configurations

**THE SCIENCE IN CONTEXT** Atoms combine in different ways in substances. Sometimes they form molecules. Sometimes they form ions. Sometimes they form metallic structures. Understanding how atoms bond together in substances enables scientists to make best use of them.

### Bad breath and smelly socks

A main cause of bad breadth is methyl mercaptan. Some bacteria in feet can also make methyl mercaptan, and make socks rather smelly.

## Molecules and ions

### Molecules

Molecules are formed from a small number of atoms joining together.

| Elements | | Compounds | |
|---|---|---|---|
| hydrogen | $H_2$ | water | $H_2O$ |
| oxygen | $O_2$ | carbon dioxide | $CO_2$ |
| nitrogen | $N_2$ | ammonia | $NH_3$ |
| phosphorus | $P_4$ | methane | $CH_4$ |
| sulfur | $S_8$ | sulfur dioxide | $SO_2$ |

**TABLE 1**: Examples of molecules.

The chemical formulae tell you how many atoms are in a **molecule**. A sulfur molecule is made from eight sulfur atoms. A methane molecule is made from one carbon atom and four hydrogen atoms.

Atoms that form a molecule are held together by covalent bonds. Two atoms share one or more pairs of electrons. Sharing one pair of electrons is the most common – it is called a single **covalent bond**.

Like atoms, a molecule has no overall electrical charge. It is neutral.

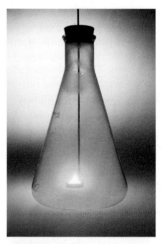

FIGURE 2: Sulfur burns in air and produces molecules of sulfur dioxide.

### Ions

When non-metals combine with metals, electrons are not shared. Instead they **transfer** from the metal atom to the non-metal atom. Both atoms are left with an electric charge. Charged atoms are called ions.

Metal ions always have a positive charge. Non-metal ions always have a negative charge. The **attraction** between these oppositely charged ions holds the compound together. This is an **ionic bond**.

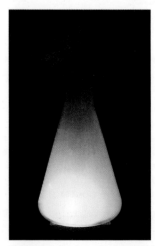

FIGURE 3: Sodium burns in chlorine and produces sodium ions, $Na^+$, and chloride ions, $Cl^-$. These bond together and make solid sodium chloride.

### QUESTIONS

1 What is a molecule?

2 Describe a covalent bond.

3 What charge will an atom have if it (a) gains an electron (b) loses an electron?



... gcse ions ... gcse molecules

# Electrons

## Stable electronic configurations

Group 0 elements are extremely unreactive. Their atoms do not share, gain or lose electrons. Their electronic configurations are stable. Atoms without stable Group 0 electronic configurations may share or transfer electrons to become stable.

## Electrons and molecules

A chlorine atom needs one more electron to have eight in its outer shell. Two chlorine atoms share a pair of electrons (one from each atom) to make a chlorine molecule ($Cl_2$). Both have a stable electronic configuration – the same as that of the Group 0 element **argon**.

An oxygen atom needs two more electrons to have eight in its outer shell. A hydrogen atom needs one more to have the same configuration as helium. Oxygen forms two covalent bonds, one with each of two hydrogen atoms. All atoms in a water molecule have a stable electronic configuration – each hydrogen the same as helium, and oxygen the same as neon.

## Electrons and ions

A chlorine atom gains an electron to get on outer shell of eight. It is now a negatively charged chloride **ion**, $Cl^-$. The electron comes from the sodium atom, leaving the second shell of the sodium atom as its outer shell. This has eight electrons, so it is stable. The sodium atom becomes a positively charged sodium ion, $Na^+$. All atoms have a stable electronic configuration – sodium the same as neon and chlorine the same as argon.

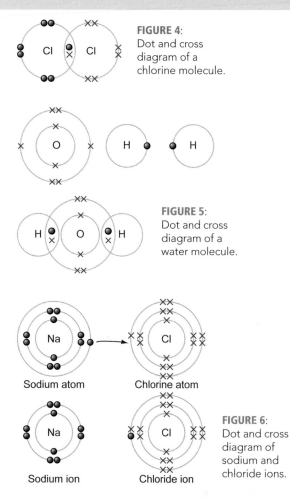

FIGURE 4:
Dot and cross diagram of a chlorine molecule.

FIGURE 5:
Dot and cross diagram of a water molecule.

Sodium atom     Chlorine atom

Sodium ion     Chloride ion

FIGURE 6:
Dot and cross diagram of sodium and chloride ions.

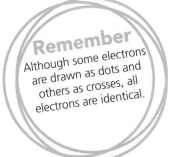

**Remember**
Although some electrons are drawn as dots and others as crosses, all electrons are identical.

## QUESTIONS

**4** What do stable electronic configurations have in common?

**5** Draw a dot and cross diagram to show the bonding in a hydrogen molecule.

**6** Why are metal ions always positively charged?

# Chemical bonds

Particles with opposite electrical charge attract one another.

In a covalent bond, the two positively charged nuclei are both attracted to the shared electrons.

In an ionic bond, positively charged ions are attracted to negatively charged ions.

## QUESTIONS

**7** A covalent bond with four shared electrons is stronger than a covalent bond with two shared electrons. Try to explain this.

# Materials from rocks

**THE SCIENCE IN CONTEXT** Earth's crust is a rich source of raw materials. Scientists, technologists and engineers combine their skills and knowledge to obtain these raw materials, using techniques such as quarrying, mining and drilling. Metal ores, limestone and fossil fuels are examples of raw materials. Some are used straight from the ground, but many are the starting materials to make useful products.

## Gold, naturally

Gold occurs in the ground as the metal itself – not as compounds. Rocks are crushed to release the tiny pieces of gold.

FIGURE 1: Panning for gold.

## Raw materials

Earth's crust provides a wealth of useful raw materials. Some are used as they are. Others need separating and purifying before they are used. However, all these substances need to be removed from the ground first. Mining and quarrying are two important methods.

Raw materials obtained by:

> quarrying include granite, gypsum, limestone, marble, sandstone, slate.

> mining include **coal**, gold, metal ores, minerals and rock salt.

### Substances used straight from the ground

Many materials are quarried or mined and used as they are.

### Social, economic and environmental impacts

Raw materials from the ground are essential to the economy. Fossil fuels such as coal are vital energy resources. Other materials are used in construction work and manufacturing various products that people use every day. Jobs are provided and money is put into the economy.

Extraction can affect the **environment**. Effects include unsightly changes to the landscape, loss of **biodiversity**, and **pollution** of soil and water. These days, in most countries, quarrying and mining companies must meet strict regulations to minimise environmental impacts.

| Raw material | Examples of uses |
| --- | --- |
| coal | fuel, making coke |
| gold | jewellery, wiring, computer chips |
| limestone and marble | construction work |
| metal ores and minerals | making metals |
| rock salt | de-icing roads |
| sulfur | making sulfuric acid and rubber |

TABLE 1: Examples of substancs used straight from the ground.

FIGURE 2: This disused slate quarry in Leicestershire was flooded to create an attractive lake.

## QUESTIONS

1 Name three substances that are (a) quarried (b) mined.

2 Give one use each for (a) limestone (b) gold (c) rock salt.

3 Suggest why obtaining raw materials from the ground is essential to the economy.

# Drilling, separating and purifying

## Drilling

Drilling wells is how crude oil (a liquid) and **natural gas** are obtained. Oil and gas are trapped in large caverns underground. Holes are drilled deep into the ground until they reach the caverns. The oil or gas is pumped to the surface.

Liquid mining is another form of drilling. The solid mineral is melted to make a liquid, which can be pumped to the surface. This is how sulfur is extracted. Superheated steam is pumped into the sulfur deposits. The sulfur melts and liquid sulfur is driven out by hot air (Figure 4).

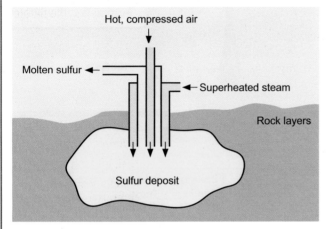

FIGURE 3: Why is superheated steam pumped into the sulfur deposit?

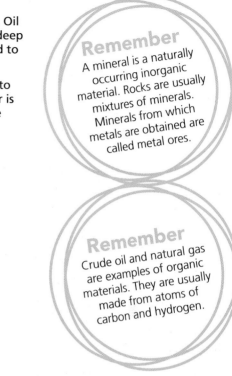

**Remember**

A mineral is a naturally occurring inorganic material. Rocks are usually mixtures of minerals. Minerals from which metals are obtained are called metal ores.

**Remember**

Crude oil and natural gas are examples of organic materials. They are usually made from atoms of carbon and hydrogen.

## Separating and purifying

Rock salt contains impurities. It is fine to put on icy roads, but not to sprinkle on your food. The mixture must be separated to produce purified salt.

The impurities are mainly clay and finely ground rock. These are not soluble in water. Salt fit to eat is obtained by:

> mixing rock salt with water – salt dissolves to give a solution, but the impurities do not

> the mixture is filtered – impurities stay on the filter paper (the residue), but the salt solution passes through and is collected (the filtrate)

> the filtrate is evaporated, leaving purified salt.

### QUESTIONS

**4** Give two examples of impurities in rock salt.

**5** Describe what happens when a mixture is filtered.

# Chemical compounds

Substances obtained from the ground are mixtures of chemical compounds. Chalk, limestone and marble are all mainly calcium carbonate ($CaCO_3$). Rock salt is mainly sodium chloride (NaCl).

### QUESTIONS

**6** Find the names of the chemical compounds in these metal ores: bauxite, haematite, malachite, zincite, galena.

### Did you know?

Mining and quarrying have a huge impact on lives, but they use less than 1% of the land in England.

Q ... gcse sulfur mining

# Crude oil

**THE SCIENCE IN CONTEXT** One hugely important resource obtained from Earth's crust is crude oil. It is transported to oil refineries, where it is separated it into useful fractions, such as petrol and diesel. Crude oil is also the raw material for the petrochemical industry, which manufactures organic compounds. The European petrochemical industry has a turnover of more than £60 billion.

## Black gold

Crude oil was dubbed 'black gold' because of the riches it brought to people who found it.

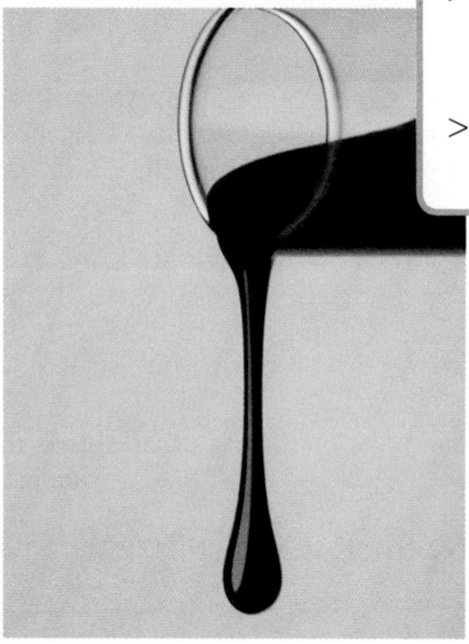

FIGURE 1: Crude oil.

## Obtaining and distilling fractions

### Crude oil is a mixture

Crude oil (sometimes called **petroleum**) is a mixture of compounds obtained by drilling. It is perhaps more accurate to talk about crude oils, since the colour, viscosity and composition varies. It depends on where the oil is drilled.

Crude oil is not much good used straight from the ground. The mixture needs to be separated into **fractions**. Examples of fractions are petrol and diesel. Fractions are obtained by distilling crude oil using **fractional distillation**.

FIGURE 2: An oil rig in the North Sea – a remarkable feat of engineering.

### Fractional distillation

Fractional distillation separates crude oil into parts or fractions. Each fraction contains a mixture of similar compounds with similar boiling points.

Compounds with larger molecules boil at higher temperatures. You don't need to remember the names of the fractions.

| Name of fraction | Boiling points (°C) | Viscosity (ease of flow) | Uses |
|---|---|---|---|
| petroleum gas | below 20 | increasing viscosity | heating; cooking; liquefied petroleum gas (LPG) fuel |
| petrol | 20 – 100 | | fuel for cars and lorries |
| naphtha | 110 – 190 | | to make other chemical compounds |
| kerosene | 170 – 240 | | fuel for aircraft; paraffin |
| diesel | 240 – 330 | | diesel fuel for cars, lorries and trains |
| fuel oil | 330 – 400 | | fuel for ships, factories and heating |
| bitumen | over 400 | | tar for road making |

## QUESTIONS

**1** Why is crude oil sometimes called black gold?

**2** Explain what is meant by 'fraction' when talking about crude oil.

**3** What is the process that is used to separate crude oil into fractions?

Q ... gcse crude oil

# Hydrocarbons and fractional distillation

## Hydrocarbons

The most common compounds in crude oil are **hydrocarbons**. These are compounds with molecules made only of hydrogen and carbon atoms – there are vast numbers of different hydrocarbons.

The compounds in a fraction are mainly hydrocarbons with similar numbers of carbon atoms in their molecules. The number of carbon atoms relates to the use of a fraction.

| Name of fraction | Carbon atoms per molecule |
|---|---|
| petroleum gas | 1 to 4 |
| petrol | 5 to 9 |
| naphtha | 6 to 10 |
| kerosene | 10 to 16 |
| diesel | 14 to 20 |
| fuel oil | 20 to 50 |
| bitumen | more than 50 |

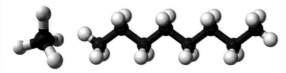

**FIGURE 3**: Models of two hydrocarbon molecules – methane and octane. How many carbon atoms are in their molecules?

## Separating fractions by fractional distillation

Distillation is done in a **fractionating column**. Crude oil is heated to about 400 °C. Most of it boils and vaporises.

> Hydrocarbons with the largest molecules and highest **boiling** points do not boil. They flow out of the bottom of the column.

> Oil vapours rise up the column, cool and condense just below their boiling point. The liquid mixture is run out of the column.

> Hydrocarbons with boiling points above: about 330 °C condense first and are run off; about 240 °C condense next and are run off; and so on (see Figure 4).

The smaller the hydrocarbon's molecules, the lower their boiling point, and the higher up the column they go before condensing.

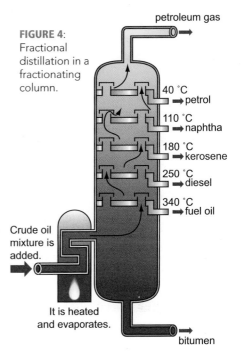

**FIGURE 4**: Fractional distillation in a fractionating column.

petroleum gas
40 °C → petrol
110 °C → naphtha
180 °C → kerosene
250 °C → diesel
340 °C → fuel oil

Crude oil mixture is added.

It is heated and evaporates.

bitumen

### QUESTIONS

**4** Explain what the term 'hydrocarbon' tells us.

**5** Methane and octane are hydrocarbons. Which is a gas and which is a liquid?

**6** Explain why petrol distils out higher up the column than diesel.

# Why do boiling points depend on size?

A compound's boiling point depends on the bonds between its molecules. The larger the molecules, the stronger the bond and the higher the boiling point. To make a liquid boil, its molecules must **absorb** enough energy to overcome the bonds – enabling them to escape from the liquid.

The table gives the boiling points of some alkanes. Alkanes are hydrocarbons with the general formula $C_nH_{2n+2}$.

| Alkane | Formula | Boiling point (°C) |
|---|---|---|
| methane | $CH_4$ | −162 |
| ethane | $C_2H_6$ | −89 |
| propane | $C_3H_8$ | −42 |
| butane | $C_4H_{10}$ | 0 |
| pentane | $C_5H_{12}$ | 36 |
| hexane | $C_6H_{14}$ | 69 |
| heptane | $C_7H_{16}$ | 98 |
| octane | $C_8H_{18}$ | 126 |

### QUESTIONS

**7** Plot a graph to show the relationships between boiling point and number of carbon atoms in an alkane. Describe the graph.

Q ... fractional distillation

# Metals from ores

**Essential notes**

- less reactive metals may be obtained from their ores by chemical reduction with carbon or carbon monoxide
- reactive metals may be obtained from their ores by electrolysis

**THE SCIENCE IN CONTEXT** Metal ores are mined from Earth's crust and are processed to make metals. Once again, scientists, technologists and engineers combine to make extraction, processing and purification possible. Techniques to obtain metals from their ores depend on the reactivity of the metal. They include chemical reduction and electrolysis.

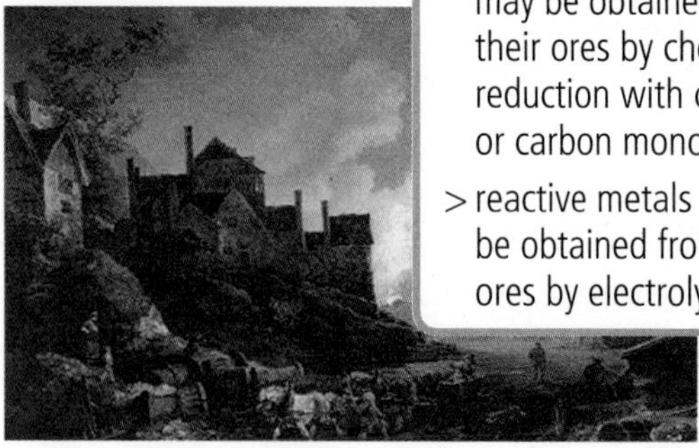

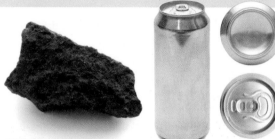

**FIGURE 1**: Darby lived in Coalbrookdale, a small village in Shropshire. Within 40 years of his discovery it became a major mining site, employing about 500 people.

## Shaping history

The blast furnace shaped history. It used coke instead of charcoal to extract iron from iron ore. Abraham Darby did it in 1709.

## Ores and extraction

### Common ores

Gold and silver are unusual. They are found in rocks as the metals themselves. Both can be mined and used straight forom the ground. Most metals occur in rocks as compounds.

Ores are rocks from which metals are obtained.

| Metal | Ore |
|---|---|
| aluminium | bauxite – mainly aluminium oxide, $Al_2O_3$ |
| iron | haematite – mainly iron oxide, $Fe_2O_3$ |
| lead | galena – mainly lead sulfide, PbS |

**TABLE 1**: Examples of common ores.

**FIGURE 2**: Aluminium has many uses, but it starts life as a piece of bauxite.

### Extracting the metal

Metals that are not very reactive are obtained by heating their oxides with coke (carbon) in a furnace. Iron **ore** is an example. The process is called **smelting**. Non-oxide ores, such as galena (a lead ore), are first converted into oxides by heating them in air. Then they are heated with coke to extract the metal.

The chemical reaction is a **reduction**. Reduction occurs when oxygen is removed from the ore to leave the metal.

Reactive metals cannot be obtained from their ores this way. They are extracted by passing an electric current through the **molten** ore. This is called electrolysis. Aluminium is an example.

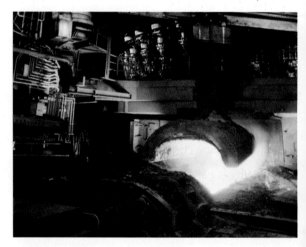

**FIGURE 3**: A modern blast furnace.

### Economics

Quarrying or mining rock, crushing it and separating the metal ore is expensive. Only rocks sufficiently rich in metal ore are usually used.

Another consideration is the energy required. Extracting metals by electrolysis is much more costly than smelting. Large amounts of energy are needed – and energy is expensive.

## QUESTIONS

1 What is a metal ore?

2 What are two main processes for obtaining metals from their ores?

3 What is removal of oxygen called?

# Extracting lead and iron

## Extracting lead

1. Lead sulfide is heated in air to make lead oxide:

   lead sulfide + oxygen (from the air)
   → lead oxide + sulfur dioxide

   $2PbS + 2O_2 \rightarrow PbO + SO_2$

2. Lead oxide is reduced by heating it with coke (a form of carbon) in a furnace. The reaction is:

   lead oxide + carbon → lead + carbon dioxide

   $2PbO + C \rightarrow 2Pb + CO_2$

3. Carbon monoxide is also used as a reducing agent:

   lead oxide + carbon monoxide → lead + carbon dioxide

   $PbO + CO \rightarrow 2Pb + CO_2$

## Extracting iron

Iron is extracted in a **blast furnace**. A number of reactions are involved. They take place in different regions of the furnace (Figure 4).

> In the yellow zone, coke (carbon) reacts with oxygen to make carbon dioxide.

   carbon + oxygen → carbon dioxide

> In the orange zone, carbon dioxide reacts with coke (carbon) to make carbon monoxide.

   carbon dioxide + carbon → carbon monoxide

> In the red zone, iron ore (iron oxide) reacts with carbon monoxide to make iron and carbon dioxide.

   carbon monoxide + iron oxide → iron + carbon dioxide

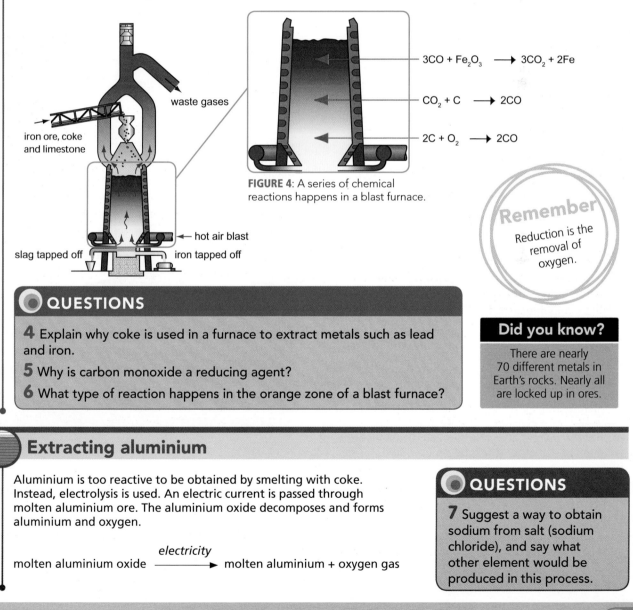

waste gases

iron ore, coke and limestone

$3CO + Fe_2O_3 \longrightarrow 3CO_2 + 2Fe$

$CO_2 + C \longrightarrow 2CO$

$2C + O_2 \longrightarrow 2CO$

**FIGURE 4**: A series of chemical reactions happens in a blast furnace.

hot air blast

slag tapped off

iron tapped off

**Remember**

Reduction is the removal of oxygen.

## QUESTIONS

**4** Explain why coke is used in a furnace to extract metals such as lead and iron.

**5** Why is carbon monoxide a reducing agent?

**6** What type of reaction happens in the orange zone of a blast furnace?

**Did you know?**

There are nearly 70 different metals in Earth's rocks. Nearly all are locked up in ores.

# Extracting aluminium

Aluminium is too reactive to be obtained by smelting with coke. Instead, electrolysis is used. An electric current is passed through molten aluminium ore. The aluminium oxide decomposes and forms aluminium and oxygen.

molten aluminium oxide $\xrightarrow{\text{electricity}}$ molten aluminium + oxygen gas

## QUESTIONS

**7** Suggest a way to obtain sodium from salt (sodium chloride), and say what other element would be produced in this process.

Q ... gcse methods of extracting metals

# Resources from the air

**THE SCIENCE IN CONTEXT** It may seem odd to think that we get raw materials from air, but we do – and they are important. Fertilisers – vital to provide the world's population with enough food – are made from nitrogen, and the nitrogen is obtained from air. Oxygen is also used in many industrial manufacturing processes.

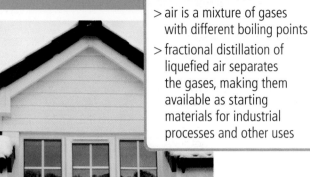

FIGURE 1: Argon fills the gap between the glass.

## Useful argon

Argon is used to fill the gap between panes of glass in double glazed windows.

## A valuable resource (Higher tier)

Air is a mixture of gases. Table 1 gives some more information about gases that are present.

Separating out the components of air provides raw materials for many industrial processes. Uses include:

> nitrogen gas to make ammonia – to make fertilisers and other products

nitrogen + hydrogen $\rightleftharpoons$ ammonia

$N_2 + 3H_2 \rightleftharpoons 2NH_3$

$\rightleftharpoons$ is used instead of $\rightarrow$ because the reaction is reversible.

> liquid nitrogen – to fast-freeze foods and biological specimens

> oxygen – in **steel** production, to help patients with breathing difficulties, in rocker propellants, and (mixed with other gases) for respiration in, for example, submarines and aircraft

> **noble gases** – argon to fill light bulbs, neon lights and signs, helium to fill balloons, and xenon headlights.

### Liquefying air

Dust is removed from air by filtering.

The dust-free air is compressed and then allowed to expand. As it expands, air cools. This is done in stages until the temperature reaches -200 °C and air has liquefied – in other words become a liquid.

At each stage a different gas separates out.

Water vapour condenses and is removed. Then carbon dioxide freezes at -79 °C and is removed. Other gases liquefy at their boiling points. The remaining mixture (mainly nitrogen and oxygen, with tiny quantities of noble gases) is separated by fractional distillation.

TABLE 1: Gases in air.

| Gas | Formula | % dry air by volume | % dry air by mass | Boiling point (°C) |
|---|---|---|---|---|
| nitrogen | $N_2$ | 78.09 | 75.47 | -196 |
| oxygen | $O_2$ | 20.95 | 23.20 | -183 |
| argon | Ar | 0.933 | 1.28 | -186 |
| carbon dioxide | $CO_2$ | 0.03 | 0.046 | -79 |
| neon | Ne | 0.0018 | 0.0012 | -246 |
| helium | He | 0.0005 | 0.00007 | -269 |
| krypton | Kr | 0.0001 | 0.0003 | -153 |
| xenon | Xe | $9 \times 10^{-6}$ | 0.00004 | -108 |

FIGURE 2: Liquid $O_2$ is delivered and stored in well-insulated vessels. You may see them in the grounds of a hospital.

## QUESTIONS

1 What is the most abundant noble gas in air?

2 Which gas in air has the lowest boiling point?

3 At what temperature does air begin to liquefy?

# Fractional distillation of liquid air (Higher tier)

The gases that make up air have different boiling points. This means that they can be separated by fractional distillation – the same process as separating crude oil, only at very much lower temperatures.

Separation takes place in a fractionating column. The column is warmer at the bottom (-185 °C) than at the top (- 190 °C). Liquefied air (at -200 °C) is fed into the bottom. Liquid nitrogen boils. Gaseous nitrogen rises to the top of the column and is piped off. Liquid oxygen remains at the bottom of the column, together with argon.

Further fractionating columns are used to separate out noble gases.

**Did you know?**

Air is the only viable source of argon, neon, krypton and xenon.

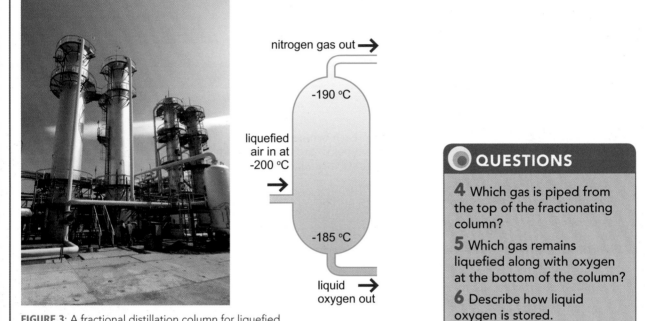

nitrogen gas out →

-190 °C

liquefied air in at -200 °C →

-185 °C

liquid oxygen out →

**FIGURE 3**: A fractional distillation column for liquefied air and a diagram showing what goes on.

**QUESTIONS**

**4** Which gas is piped from the top of the fractionating column?

**5** Which gas remains liquefied along with oxygen at the bottom of the column?

**6** Describe how liquid oxygen is stored.

# Boiling points and size (Higher tier)

Compounds with larger molecules have higher boiling points than compounds with smaller molecules.

The elements in Group 0 are gases. They exist as atoms. The diameters of their atoms are given in Table 2.

**Did you know?**

A nanometre (nm) is $10^{-9}$ metres, or one-sixth of a millimetre.

| Element | Diameter of atom (nm) |
|---------|----------------------|
| helium | 0.062 |
| neon | 0.076 |
| argon | 0.142 |
| krypton | 0.176 |
| xenon | 0.216 |

**TABLE 2**: Diameters of some Group 0 gases.

**QUESTIONS**

**7** Plot a graph of boiling point against diameter of atom. Describe the shape of the graph.

Q ... fractional distillation of air

# Chemical equations

## Essential notes

> chemical symbols and formulae are used to write equations

> there must the same number and types of atoms in the products as there were in the reactants

**THE SCIENCE IN CONTEXT** Knowing about amounts of reactants and products in chemical reactions is important when making products (both in laboratories and manufacturing plants). It is also essential in chemical analysis, which is used, for example, to test the purity of reactants and products. Chemical equations provide preparative chemists, analytical chemists and manufacturers with vital information about reacting quantities.

### The first chemical equation

Jean Beguin wrote the first chemical equation in 1615. It did not use symbols, just words. It is not easy to follow and looks quite unlike the equations that scientists write today.

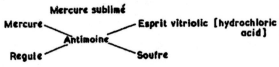

FIGURE 1: Beguin's equation. Type the French words into a search engine and draw a translated equation.

## Chemists' shorthand

### Symbols, formulae and equations

Each element has a name and a unique symbol. Each compound has a name and a formula. The formula tells you the types of atoms it was made from and their relative numbers. You can use these formulae to describe chemical reactions in 'chemistry shorthand'.

Word equations use the names of **reactants** and products. For example:

lead oxide + carbon ➡ lead + carbon dioxide

### In every chemical equation:

> reactants are on the left of the arrow.

> products are on the right of the arrow.

> + means 'reacts with' or 'and'

> ➡ means 'go to'.

However, word equations do not tell you anything about the numbers of atoms involved. For this, symbol equations are used.

### Writing a symbol equation

To write a symbol equation, replace names with chemical formulae. Using the example above:

$$2PbO + C \rightarrow 2Pb + CO_2$$

You may find the coloured circles method useful to track the atoms:

The equation is balanced because the types and numbers of atoms are the same on both sides of the arrow. No atoms disappear. No new atoms are created. They are just reorganised.

Numbers in front of formulae show how many 'formula units' of a substance react or are produced. No number means 'one formula unit'. A formula unit is the way a chemical formula is written. So, PbO means one formula unit. 2PbO means two formula units. $3O_2$ means three formula units.

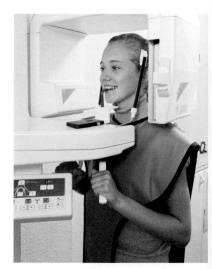

FIGURE 2: The reaction between lead oxide and carbon can be used to obtain lead from its ores. Lead has many uses. The apron in the photograph is lined with lead, to protect the wearer from harmful radiation while having a dental X-ray.

## QUESTIONS

**1** What is the name for substances formed during a chemical reaction?

**2** What is the advantage in using chemical equations instead of words?

Q … gcse chemical equations

# Balancing equations (Higher tier)

The word equation for the reaction of methane and oxygen is:

methane + oxygen → carbon dioxide + water

Replacing names with chemical formulae gives:

$CH_4 + O_2$ → $CO_2 + H_2O$

However, there are:

> one C atom, four H atoms and two O atoms on the left of the arrow

> one C atom, two H atoms and three O atoms on the right of the arrow.

Now, two H atoms cannot be lost and an extra O atom cannot be created.

You cannot change the formulae, only the numbers of formulae units. If you put 2 in front of $O_2$ and 2 in front of $H_2O$ the equation is balanced:

$CH_4 + 2O_2$ → $CO_2 + 2H_2O$

You can use coloured circles to track atoms:

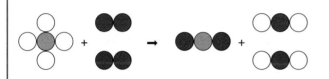

FIGURE 3: Methane burns to make carbon dioxide and steam.

| State | State symbol |
|---|---|
| solid | s |
| liquid | ℓ |
| gas | g |
| aqueous (dissolved in water) | aq |

## State symbols

In the reaction between methane and oxygen, water is produced as steam – a gas.

$CH_4 + 2O_2$ → $CO_2 + 2H_2O$

However, the equation does not tell you this. This is why state symbols are sometimes used in chemical equations.

The symbol is put in brackets after the formula. The reaction between methane and oxygen is written:

$CH_4(g) + 2O_2(g)$ → $CO_2(g) + 2H_2O(ℓ)$

## QUESTIONS

**3** What do equations tell us, besides which substances are involved in the reaction?

**4** How do equations show the number of formula units reacting?

**5** Why do equations need to be balanced?

# More complex equations (Higher tier)

Sodium carbonate solution reacts with hydrochloric acid to make sodium chloride, water and carbon dioxide. Using symbols give:

$Na_2CO_3 + HCl$ → $NaCl + H_2O + CO_2$

There are two Na atoms in the reactants, but only one in the products. 2 in front of NaCl is needed:

$Na_2CO_3 + HCl$ → $2NaCl + H_2O + CO_2$

Now, there are two H atoms in the products, but only one in the reactants. 2 in front of HCl is needed:

$Na_2CO_3 + 2HCl$ → $2NaCl + H_2O + CO_2$

There are six O atoms on each side of the arrow.

The equation is balanced.

### Did you know?

Berzelius introduced the system of chemical symbols still in use. That was 200 years ago, but they were hated by chemists of the time.

## QUESTIONS

**6** Balance the equations
(a) $N_2 + H_2$ → $NH_3$
(b) $Na + H_2O$ → $NaOH + H_2$

# Making products

**THE SCIENCE IN CONTEXT** Companies need to make profits, to repay investors and to invest in future projects. Manufacturing processes are designed to give the maximum product for the minimum cost, as safely and quickly as possible, and with minimum impact on the environment. As well as the reactants, manufacturing costs include energy and waste management.

## Soapless detergents

Soapless detergents are manufactured from crude oil by a series of chemical processes.

**FIGURE 1**: Traditionally, soap is made by boiling oils and fats with an alkali.

## Production, safety, economics and environment

### Production

Virtually everything you use has been manufactured from raw materials. Earth's crust, sea and atmosphere, and living organisms provide these raw materials.

Processes to make these products must:

> be safe

> be economical

> have minimum environmental impact

> use the correct amounts of materials.

### Safety

A hazard is a substance or action that could cause harm. Risk is the likelihood of harm being done. The hazards of materials and processes must be identified and risks from using them minimised.

### Economics

Costs of manufacture include:

> raw materials and other substances that are needed in the process

> energy consumption

> waste disposal.

Products make money for companies that make them, but they must offer 'value for money' to the consumer.

### Environment

Building a manufacturing plant could affect the environment. So could running it and disposing of waste. Building and operating must be done in accordance with many rules and regulations.

### Did you know?

The UK chemical industry employs 180 000 people working for 3000 organisations.

**FIGURE 2**: Hazard symbols are clearly displayed where substances are being processed.

**FIGURE 3**: Factories require extensive planning and consultation before being built.

## QUESTIONS

**1** Name three important areas to consider when making products.

**2** What is the difference between hazard and risk?

**3** Name three things that contribute to the cost of making a product.

# Calculating masses (Higher tier)

Using the right quantities of materials to make a product is important. Imagine extracting lead from lead oxide by heating it with carbon. If insufficient carbon is used you would end up with a mixture of lead and lead oxide – not very helpful. How can the necessary quantities be calculated?

Chemical equations must be balanced. The total mass of products must equal the total mass of the reactants. This is the conservation of matter.

*Example 1*

It is known that 48 g of magnesium reacts with 32 g of oxygen to make 80 g of magnesium oxide.

The ratio is always: 48 : 32 : 80

   magnesium    oxygen    magnesium oxide

Using simple proportions you can work out that to make:

> 8 g of magnesium oxide needs 4.8 g of magnesium with 3.2 g of oxygen

> 160 g of magnesium oxide needs 96 g of magnesium with 64 g of oxygen.

You can also work out that, if you have:

> 12 g of magnesium, you need 8 g of oxygen to react with it and 20 g of magnesium oxide would be made.

*Example 2*

Suppose you have 223 g of lead oxide and you want to work out what mass of carbon is needed to reduce it to lead.

To start you need to know that:

446 g of lead oxide reacts with 12 g of carbon. Although you do not need it for this calculation, 414 g of lead and 44 g of carbon dioxide are formed.

So, 223 g of lead needs 6 g of carbon.

**FIGURE 4**: What mass of magnesium oxide is made by burning 2.4 g of magnesium in 1.6 g of oxygen?

## QUESTIONS

**4** What mass of magnesium is needed to make 10 g of magnesium oxide?

**5** What mass of lead can be obtained from 8.92 g of lead oxide?

# Recovering metals from waste (Higher tier)

In the past, land and water may have been contaminated by waste from disused mines and former industrial sites. One type of contamination involves compounds of toxic metals, such as cadmium, nickel and cobalt. They need to be removed if the land or water is to be used.

A recent approach is to use plants to remove these compounds. The plants absorb metal compounds through their roots. They can then be burned and metals recovered from the ash. This is called **phytomining**. Not only does it decontaminate the soil, it is also enables metals, some of which may be very expensive, to be recovered.

**FIGURE 5**: Plants such as this water hyacinth can absorb metal compounds from contaminated land and water.

## QUESTIONS

**6** How do you think a metal might be recovered from the plants that absorbed its compounds?

# Preparing for assessment: Applying your knowledge

*To achieve a good grade in science, you not only have to know and understand scientific ideas, but you need to be able to apply them to other situations and investigations. These tasks will support you in developing these skills.*

## ✺ New material discovered

It is 1791 in Manaccan, a small village in Cornwall. The local vicar is the Reverend William Gregor. His hobby is collecting and studying samples of rocks and minerals. He has already collected many local samples and catalogued them, but one day he comes across something he has not seen before. He shows his wife, Charlotte, a reddish brown substance. "You remember the black sand I found down in the valley?" he explains hurriedly, "You know, that turned out to be magnetic? I do believe it is a new material."

Gregor analysed the sand. He concluded that it consisted of two metal oxides – iron oxide and a white metallic oxide that he could not identify. He called this unknown metal 'manaccanite' after the village. Although he did not succeed in purifying it, his findings were published that year, but aroused little interest.

## ✺ TASK 1

(a) Gregor's 'black sand' was sand mixed with a mineral that is now called ilmenite. It is an oxide of iron and another metal (that Gregor called manaccanite). Describe how metals are obtained from their oxide ores.

(b) Two metals can be obtained from ilmenite. One was what Gregor called manaccanite. What is the other metal?

(c) Explain why Gregor's 'black sand' is a mixture of compounds, but the metals that can be obtained from them are elements.

## ✺ TASK 2

(a) Describe where metals are found in the periodic table.

(b) Manaccanite has a relatively low density. Unlike Group 1 and Group 2 metals, it does not react with water or steam. Narrow down its likely place in the periodic table.

(c) Manaccanite has a similar strength to steel, but is 40% lighter. It is denser than aluminium, but is significantly stronger. Its combination of high strength and low density means that it is widely used in aircraft construction – around 77 tonnes are used to build an Airbus A380 'double-decker' airliner. What is the modern name for the manaccanite and what is its symbol?

## ✸ TASK 3

(a) Atoms of manaccanite have 22 protons. What is its atomic number and how many electrons are there in each atom?

(b) The most common isotope of manaccanite has a mass number of 48. How many neutrons are there in an atom of this isotope?

## ✸ TASK 4

(a) Manaccanite cannot be extracted from its oxide ores by heating them with carbon. Instead, its ores are heated with carbon and chlorine to make the chloride of manaccanite ($XCl_4$) and carbon monoxide (CO). One of its ores, called rutile, has the formula $XO_2$. Write a word equation and a balanced symbol equation for the reaction of rutile with chlorine and carbon. Note: Instead of writing $XO_2$ and $XCl_4$ use the correct chemical symbol for manaccanite.

(b) $XCl_4$ is heated with sodium or magnesium to extract metal X. Write a word equation and a balanced symbol equation for the reaction of $XCl_4$ with (a) sodium, (b) magnesium. Note: Instead of writing $XO_2$ and $XCl_4$ use the correct chemical symbol for manaccanite.

## ✸ MAXIMISE YOUR GRADE

| Answer includes showing that you can... |
| --- |

**E**

classify materials, for example, iron as an element, iron oxide as a compound and black sand as a mixture.

describe how metals are separated from their ores, for example, by reducing iron oxide with carbon.

describe the structure of an atom in terms of the number of protons, neutrons and electrons and their arrangement.

**C**

use balanced chemical equations to represent the formation of products in a chemical reaction.

define atomic number and mass number.

describe the reaction between metal ores and carbon as a reduction reaction.

**A**

calculate the number of each sub-atomic particle in an atom from its atomic number and mass number.

balance chemical equations.

# Unit 1 Theme 1 Checklist

## To achieve your forecast grade in the exam you will need to revise

Use this checklist to see what you can do now. Refer back to the relevant pages in this book if you are not sure. Look across the three columns to see how you can progress. **Bold** text means Higher tier only.

Remember that you will need to be able to use these ideas in various ways, such as:

> interpreting pictures, diagrams and graphs

> applying ideas to new situations

> explaining ethical implications

> suggesting some benefits and risks to society

> drawing conclusions from evidence that you are given.

Look at pages 250–271 for more information about exams and how you will be assessed.

| To aim for a grade E | To aim for a grade C | To aim for a grade A |
|---|---|---|
| Know that Earth, the solar system and the Milky Way are part of the universe. | Know that Earth is part of the solar system, which is part of the galaxy called the Milky Way. | Discuss the implications of the position of Earth in the solar system. |
| Know that observations of the solar system and galaxies can be carried out on Earth or from space. | Understand that observations are made with telescopes that detect electromagnetic radiations such as visible light, radio waves or X-rays from space. | Explain how observations provide evidence for changes taking place in the universe. |
| Recall that red-shift provides evidence that the universe is expanding and supports the Big Bang theory. | Explain how red-shift provides evidence that the universe is expanding. | Explain how red-shift supports the Big Bang theory. |
| Describe how, if a wave source is moving relative to an observer, there will be a change in the observed wavelength and frequency (Doppler effect). | Explain why there is a red-shift in light observed from most distant stars and galaxies. | Explain why the further away that stars or galaxies are, the more their light is red-shifted. |
| Recall that Earth consists of a mantle, core and crust, surrounded by the atmosphere. | Know that the surface of Earth has changed over time as a result of cooling. | Know that Earth's crust and the upper part of the mantle are cracked into a number of large pieces called tectonic plates. |
| Describe how convection currents within the mantle cause the movement of tectonic plates. | Explain how movement of tectonic plates can cause earthquakes and volcanoes. | Discuss the accurate prediction of earthquakes and volcanic eruptions. |
| Know that, during the first billion years of Earth's existence, there was intense volcanic activity, which released the gases that formed the early atmosphere and water vapour that condensed to form the oceans. | Know that some theories suggest that, during the first billion years, Earth's atmosphere was mainly carbon dioxide with little or no oxygen. | **Know that there may have been water vapour and small amounts of methane, hydrogen and ammonia in Earth's early atmosphere.** |

## To aim for a grade E

Describe how Earth's atmosphere allows light energy radiated from the Sun to pass through.

Describe how plants and algae produced the oxygen, found in today's atmosphere, by photosynthesis.

Know that each element has a symbol and that each compound has a formula.

Describe the structure of the atom in terms of numbers of protons, neutrons and electrons, their arrangement and charge.

Describe atoms, molecules and ions.

Know that useful materials can be removed from the ground by mining or quarrying.

Give examples of substances used straight from the ground.

Explain why mass is conserved in chemical reactions and that, during a reaction, products with different properties are formed as a result of atoms rearranging.

Discuss the social, economic and environmental impacts of exploiting Earth's crust, sea, atmosphere and living organisms.

Explain why, in order to produce a product economically and safely, it is important that the correct amount of material is used.

## To aim for a grade C

Know that greenhouse gases in the atmosphere keep temperatures on Earth stable and warm.

Distinguish elements and compounds.

Distinguish compounds and mixtures.

Define the terms atomic number and mass number.

Explain the difference between atoms, molecules and ions.

Describe how salt is separated from rock salt before use.

Describe how fuels (hydrocarbons) are separated from crude oil by fractional distillation.

Know that, when producing new products, chemical reactions can be represented by using balanced chemical equations.

Interpret symbol equations in terms of numbers of atoms.

Calculate the mass of reactant or product from information given about the other substances in an equation.

## To aim for a grade A

Explain how greenhouse gases in the atmosphere keep temperatures on Earth stable and warm to support life by allowing short-wave radiation to pass through the atmosphere to Earth's surface but absorb the outgoing long-wave radiation.

Discuss the implications of changes to the composition of the atmosphere over time.

Calculate the number of protons, neutrons and electrons in an atom of an element given the atomic number and mass number of the element.

Describe how reactive metals are separated from their ores by electrolysis of molten compounds and less reactive metals by using carbon and carbon monoxide as reducing agents.

**Describe air as a mixture of gases with different boiling points that can be fractionally distilled to provide new materials for industrial processes.**

**Balance chemical equations.**

Describe methods of cleansing coal and metal mines such as phytomining.

# Unit 1 Theme 2: Life on our planet

## What you should know

### Variation

Scientists use classification to sort living organisms into groups, including the plant and animal kingdoms.

Plants and animals vary owing to genetic or environmental factors.

- Name the group of animals that you belong to.

### Feeding relationships

Plants use photosynthesis to make food.

Food chains show what eats what in a community.

- Where does the energy come from at the beginning of a food chain?

### The carbon cycle

Carbon dioxide is a gas made from carbon and oxygen.

Carbon dioxide is constantly being recycled in the environment.

- Where is carbon dioxide found?

# You will find out

## Life on our planet

> Animals and plants can be classified using their physical characteristics.

> Microbes, animals and plants have special adaptations to allow them to survive in the conditions in which they normally live.

> Evolution takes place through natural selection.

> Plant growth is affected by light, temperature, day length and gravity, and is controlled by auxins.

## Biomass and energy flow through the biosphere

> Green plants and algae use sunlight for photosynthesis to make carbohydrates, fats and proteins, using the carbon from carbon dioxide.

> Food chains show the flow of matter and energy between producers and consumers.

## The importance of carbon

> Carbon is cycled through the air, ground, plants, animals and fossil fuels.

> Processes in the carbon cycle include photosynthesis, respiration, feeding, burning and limestone formation.

> Human activity affects the natural balance of the carbon cycle.

# Variety

**THE SCIENCE IN CONTEXT** Life on Earth has slowly changed (evolved) over millions and millions of years. It is still changing. The variety of life is huge. To help them name, identify and share information about organisms, scientists classify this wide variety of life into different groups. Taxonomists (scientists who classify species) are rare, but their work is vital.

## The oldest fossils

The oldest known fossils are Australian stromatolites. Over three billion years old, these rocks were built up by photosynthetic bacteria.

**FIGURE 1**: Australian stromatolites.

### Essential notes

> life can be categorised into kingdoms

> animals and plants can be classified using their physical characteristics

> classification is an international method of grouping organisms to help name and identify them

## Evolution and classification

Fossil records show how organisms evolved. The evolution from very simple organisms to very complex ones took millions and millions of years.

Classification means grouping organisms according to their physical characteristics. By using an international method of classification, scientists can name, identify and share information about organisms.

Organisms are grouped into kingdoms such as:

> Microbes – can be seen only by using a microscope

> Plants – usually have roots and leaves and can carry out photosynthesis

> Animals – cannot make their own food, can move about.

Each kingdom has a basic body plan that is different from the other kingdoms.

Organisms in a kingdom can be separated into smaller groups and, then again into even smaller groups. Altogether there are seven levels of classification – starting with kingdom and ending up at species.

For example:

| Classification | You | Domestic cat |
|---|---|---|
| Kingdom | animal | animal |
| Phylum | chordate | chordate |
| Class | mammal | mammal |
| Order | primate | carnivore |
| Family | hominid | felid |
| Genus | *Homo* | *Felis* |
| Species | *Homo sapiens* | *Felis catus* |

You do not neeed to learn these terms, but it shows how levels of classification narrow down.

**FIGURE 2**: Fossils are the preserved remains or traces of things that lived way back in the past.

### Did you know?

Between 2006 and 2009, scientists across the world discovered 114 000 new species.

### QUESTIONS

**1** What is meant by classification?

**2** Why do scientists classify organisms?

**3** What kinds of characteristics are used to classify organisms?

🔍 ... evolution

# Identification keys

You can use physical characteristics to identify the groups to which an organism belongs. One type of key uses paired statements. You identify one characteristic at a time until you have identified the organism. Figure 3 shows how a paired statement key may be used.

## Species

Organisms belonging to the same species:

> look similar to each other

> can interbreed to give fertile offspring (meaning they also interbreed).

Donkeys and horses can interbreed, but their offspring (a mule) cannot breed. So, donkeys and horses do not belong to the same species.

Almost two million species have been discovered and named. It is not known how many are yet to be discovered, but it's estimated there are between five and thirty million species alive today.

### QUESTIONS

**4** Explain how a key is used to classify organisms.

**5** What makes one species different to any other?

**Paired statement key for garden invertebrates:**

1. legs – go to 2
   no legs – go to 3

2. six legs – go to 4
   eight legs – spider

3. shell – snail
   no shell – go to 5

4. coloured wings – butterfly
   colourless wings – fly

5. segments – earthworm
   no segments – slug

**Branched key for garden invertebrates:**

FIGURE 3: Paired statement key and branched key example.

# Avoiding confusion

How do scientists avoid confusing so many different species? Everyday names are not helpful. For example, 'buttercup' is used for different species, including field buttercup, creeping buttercup and bulbous buttercup.

Internationally agreed, double scientific names based on Latin provide the answer. The first name (called the genus) is like a surname and is used for closely related species.

Domestic cats, for example, are called *Felis catus* and wildcats are *Felis silvestris*.

The plant in Figure 4 has numerous common names, such as Jack in the pulpit, cuckoo pint, wild arum, lords and ladies, devils and angels, cows and bulls, Adam and eve, bobbins, naked boys, starch root and wake robin. It has only one internationally agreed scientific name: *Arum maculatum*.

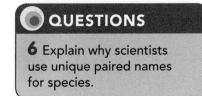

FIGURE 4: *Arum maculatum*.

### QUESTIONS

**6** Explain why scientists use unique paired names for species.

# Survival and adaptation

**THE SCIENCE IN CONTEXT** Adapt or die is the harsh reality of life on Earth. To survive, plants, animals and microbes adapt to their natural environments. Since conditions on Earth are continually changing, the characteristic features of an organism may change. It evolves and adapts to environmental changes. Biologists, ecologists and environmental scientists study these processes.

## Living the extreme

Extremophilic microbes have been found around deep sea volcanic vents in water at 400 °C and in the Antarctic at -20 °C. In the deepest part of the Pacific Ocean some microbes survive at pressures more than 1000 times greater than at the surface.

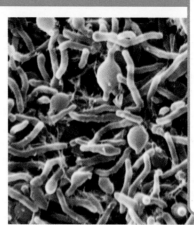

**FIGURE 1**: Some microbes can survive in extremely high radiation, acidity or salt concentrations.

## Survival

All organisms need different things to survive. You need oxygen (from the air), water and food. You also need the temperature to be neither too high nor too low.

Organisms obtain what they need to survive from their surroundings and from other living organisms.

> Plants need water, nutrients, and sunlight and carbon dioxide (to photosynthesise).

> Animals need food, mates, shelter and a suitable territory.

In order to survive, organisms **adapt** to the conditions in which they normally live. **Adaptations** include:

> size and **surface area** – for example, small or large animals, large and small leaves

> water intake and storage – for example, roots and water storage tissues

> insulation – for example, fur and fat.

A species can be identified from its physical characteristics.

Some of these characteristics enable organisms to survive in the conditions in which they live.

### Did you know?

Extremophilic means 'loving the extreme'. The existence of extremophilic microbes (extremophiles) leads scientists to believe that there could be life on other planets in the solar system despite the harsh conditions found there.

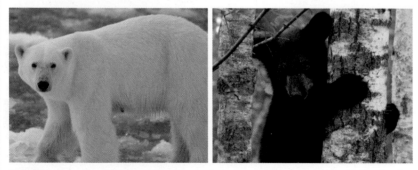

**FIGURE 2**: Bears belong to the same species, but polar bears are adapted to live in cold, icy conditions and black bears to live in hot sunny environments.

### ⬤ QUESTIONS

**1** Name three things that a plant needs to survive.

**2** Name three adaptations of animals that help them survive.

**3** Describe a difference between polar bears and black bears that allows them to live in different environments.

Q ... gcse adaptation

# Different environments

## Hot environments

In hot **environments**, organisms must keep cool enough to survive:

> Plants evaporate water from their leaves.

> Animals can **sweat**, which then evaporates.

**Evaporation** is the change of water from liquid to vapour. Energy is transferred from the organism to the water to make it evaporate, cooling the surface of the organism. Large leaves provide a larger surface area for evaporation to happen.

## Dry environments

In dry environments, such as deserts, organisms need to conserve water.

Cacti, for example, have:

> spines rather than big leaves and a thick waxy surface to reduce evaporation losses

> large cells in swollen stems to store water, and shallow roots to absorb rain water.

Camels, for example, have:

> a hump which contains fat that can be broken down in respiration to release water

> a large stomach so that they can drink large amounts of water when it is available.

## Cold environments

In cold environments, such as the Arctic, organisms need to keep warm.

> Plants such as the Arctic willow are small and grow close together and close to the ground. This protects them from the cold and strong winds. They also have shallow roots because only the surface of the soil thaws.

> Animals like polar bears have thick hairy coats and thick layers of fat under their skin for increased insulation.

**FIGURE 3**: How is the camel adapted to survive in deserts?

**FIGURE 4**: How is the Arctic willow adapted to survive in cold conditions?

### ◉ QUESTIONS

**4** Explain why some plants have very deep roots.

**5** Explain how you can survive in a temperature higher than your body temperature.

# Energy, size and surface area

Animals that live where it is very cold tend to grow larger than animals that live where it is very hot.

Large animals have a lower surface area to volume ratio than small animals. Therefore, they transfer energy to their surroundings.

Animals living in very hot environments often have large ears. A high surface area to volume ratio increases the **rate of energy transfer** to the surroundings. This helps to keep them cool.

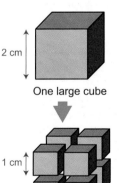

2 cm

One large cube

1 cm

Eight smaller cubes

**FIGURE 5**: Breaking an object into smaller pieces increases its surface area. What is the ratio of surface area to volume for (a) the single cube (b) the eight smaller cubes?

### ◉ QUESTIONS

**6** Hypothermia is caused by excessive heat loss, causing a dangerous drop in body temperature. Explain why babies are more prone to hypothermia than adults.

# Natural selection

**THE SCIENCE IN CONTEXT** When their living conditions change for the worse, only the better-adapted members of a species survive. They have certain characteristics that help them to do this. Less well adapted individuals perish in the new conditions, while those that are better adapted pass the characteristics that enabled them to survive to their offspring. This is natural selection. So the species evolves.

FIGURE 1: Dogs and wolves are the same species.

## Dog breeds

There are more than 600 dog breeds. Despite widely varying characteristics, all dog breeds are still the same species as the wolf.

## Natural selection theories

Darwin proposed his theory of **natural selection** in 1859. He believed that individuals that were better adapted would survive longer and have more offspring. Characteristics that adapted organisms to their surroundings would be selected for and build up gradually over many years. Eventually this gives new species different from the original ones.

Darwin did not know about **genes**. However, the modern theory of natural selection is based on understanding them.

The modern theory says that:

> individual members of a species show wide **variation** because of differences in their genes

> some individuals will be better adapted to the conditions in their environment – they are more likely to survive to breed successfully

> the genes for the useful adaptations will be passed on to the next generation.

**Remember**
Species evolve to adapt better to environmental conditions. Less well adapted species will die out.

**Did you know?**

Charles Darwin was the first to use the term 'natural selection'. He wrote in *On the Origin of Species*, "I have called this principle, by which each slight variation, if useful, is preserved, by the term Natural Selection."

FIGURE 2: In modern theories, how is the variation between individual members of a species explained?

## QUESTIONS

**1** What is the link between natural selection and survival?

**2** Use Darwin's theory to explain how giraffes could evolve from a short-necked ancestor.

**3** In the modern theory, what passes useful adaptations on to the next generation?

## Darwin's finches

Comparing the differences and similarities between species helps scientists to understand the relationship between evolution and environmental conditions.

In 1831 Darwin set off on a five-year expedition on board HMS Beagle. He found unusual species on the isolated Galapagos Islands, in the Pacific Ocean, that helped him develop his ideas laters. These included several species of finches: each species had its own characteristic beak.

| Ground finch | Cactus finch | Vegetarian tree finch | Insectivorous tree finch | Warbler finch | Woodpecker finch |
|---|---|---|---|---|---|
|  |  |  |  |  |  |
| eats seeds which it crushes with its strong beak | long, slender beak sucks up nectar from cactus flowers | curved beak, like a parrot, eats buds and fruits | stubby beak to eat beetles and other insects | small beak to catch little insects and spiders from leaves and twigs | uses a cactus spine to prize small insects out from cracks in tree bark |

**FIGURE 3**: Finches on the Galapagos Islands.

Darwin's finches all shared the characteristics of other species of finches. Could they have evolved from a common ancestral finch species that found its way to the islands a long time ago?

When species have a common ancestor but different features, these features have developed to enable the species to adapt to different environmental conditions. The finches had different beaks to eat different foods.

When species share features, but have evolved from different ancestors, these features are adaptations to suit similar conditions. Away from the Galapagos, birds which are not finches have beaks like them.

### QUESTIONS

**4** Explain why birds that are not finches can have beaks like Darwin's finches.

## Mutations

Mutations alter the code for a gene. They may cause an individual to be less well adapted to their environment. If they are better adapted, they are more likely to survive and pass on the new form of the gene. Mutations can lead to more rapid change in a species.

### Evolution of a superbug

Bacteria mutate easily. *Staphylococcus* is a group of bacteria commonly found in peoples' noses and throats. **MRSA** stands for methicillin-resistant *Staphylococcus aureus*. These are strains of bacteria which are **antibiotic** resistant. They are much harder to treat, needing larger doses of more powerful antibiotics. They are more likely to cause serious illness or death.

When you take an antibiotic, the weaker strains of *Staphylococcus* will be killed, but the naturally resistant ones are more likely to survive. Then, the next time you have an infection, it is much more likely to be caused by a **resistant strain**.

### QUESTIONS

**5** Suggest why you are much more likely to get an MRSA infection in hospital than at home.

# Plant growth

## THE SCIENCE IN CONTEXT

Plants are invaluable sources of food and other useful materials. They need the right conditions to develop and grow healthily. Understanding what plants need and being able to provide these conditions is an important area of scientific research, undertaken by plant biologists and agricultural scientists, for example. Professional growers such as horticulturists and farmers also benefit from this understanding.

### Giant redwoods

Giant redwods have fibrous bark nearly one metre thick, providing fire protection. The seeds germinate in soil in full sunlight – they grow where periodic wildfires clear the competing vegetation.

FIGURE 1: Giant redwoods rely on fire to eliminate the competition.

## Habitats and survival

All organisms live in a **habitat**. It supplies the materials needed to survive, grow and reproduce. This includes using other organisms that live there.

Any habitat, such as a field, pond, woodland or seashore, will have a limited supply of materials. Organisms that live there **compete** with each other to get what they need.

A habitat usually has many plants growing in it. The plants compete with each other for light, water and nutrients in the soil.

> Light needed for photosynthesis is absorbed by leaves. The energy is used to make the plant's food. Taller plants or plants with larger leaves compete better for the available sunlight, but may lose water more easily or be more susceptible to wind damage.

> Water is absorbed through roots. Plants may have shallow or deep root systems.

> Plants also need a range of mineral salts dissolved in water in the soil, such as nitrates for making proteins.

FIGURE 2: Different plants grow in this vegetable garden – some wanted, some not.

# Plant growth and development

Plants cannot move. Instead they have to control their growth and development to make the most of their surroundings. Plants can detect external factors. They grow towards the sunlight, water and nutrients that they need to survive. They also control their life cycles, so that seeds germinate and flowers grow in the best conditions possible.

Some plants such as carnations flower only when nights are short. They are called long-day flowering plants. Other plants such as chrysanthemums flower as nights grow longer. These are short-day flowering plants.

Seeds need both water and warmth to germinate, and to grow shoots and roots. Warmer temperatures allow enzymes to convert stored food such as starch to soluble sugars. These can then move to the growing shoot and root.

Plants use plant **hormones** called auxins to grow. Auxins cause cells to grow in shoots but inhibit growth in root cells.

> Shoots grow towards light (**phototropism**). Light on the side of a shoot causes auxins to concentrate on the shady side; cells grow larger there and shoots bend towards the light.

> Roots grow in the direction of gravity (**gravitropism**). Gravity causes auxins to accumulate on the bottom side of roots, inhibiting cell growth. Cells grow larger on their top side and roots bend downwards and grow into the soil.

FIGURE 3: Why do plants grow towards light?

FIGURE 4: What would happen to the roots of a plant if it was grown in space?

## QUESTIONS

**4** In what seasons would you expect carnations and chrysanthemums to flower?

**5** Why do shoots need to grow towards light and roots into the soil?

# Succession

The UK used to be covered largely in forest – but not any more. Environments change because there is a **natural succession**. Species change the environment and new species better adapted to the new conditions take over. Rock is colonised by lichens which die and form thin soil. Grasses take over. The soil improves and, after a succession of other plants, eventually trees grow and the succession ends.

FIGURE 5: Part of the New Forest.

## QUESTIONS

**6** Suggest why the UK has so little woodland and forest today.

# Harnessing sunlight

**THE SCIENCE IN CONTEXT** To survive, all organisms need supplies of energy stores, in other words food. Certain organisms can harness sunlight and use it to make glucose, and other organic compounds, from carbon dioxide and water. This is photosynthesis. Glucose is the source of energy used in cells to make chemical reactions happen. Organisms that cannot photosynthesise get their energy supplies from those that can by feeding on them.

## Sunlight

Earth receives more energy from the Sun in one hour than the whole world's population uses in an entire year.

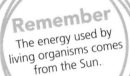

**FIGURE 1:** The Sun.

## The biosphere and biomass

Earth's **biosphere** is all the organisms on Earth and that part of Earth that they live in.

Sunlight passes through the atmosphere and enters the biosphere, carrying energy from the Sun to Earth. Organisms must capture and store this energy. They use it to carry out all the processes needed to stay alive.

Green plants and algae are able to use sunlight in **photosynthesis**. Photosynthesis changes carbon dioxide and water into glucose, an **organic compound**, and oxygen gas.

$$\text{carbon dioxide + water} \xrightarrow{\text{sunlight}} \text{glucose + oxygen}$$

Organic compounds are made from atoms of carbon and hydrogen. Some of them are made from carbon, hydrogen and atoms of other elements – mainly oxygen and nitrogen. Glucose, for example, is made from atoms of carbon, hydrogen and oxygen.

Organic compounds that make up the bodies of organisms are called **biomass**. Energy that came from the Sun and was captured in photosynthesis is stored in biomass.

Energy is made available for use by organisms during **respiration**. Energy stored in organic compounds is released – it keeps organisms alive and allows them to grow and reproduce.

**FIGURE 2**: Why are green plants important to us?

**Remember**
The energy used by living organisms comes from the Sun.

### Did you know?

Farmers grow crops so that the crops' biomass can be used as a renewable source of energy. However, the use of crops for fuel has caused food crops to rise in price.

### QUESTIONS

**1** Which process converts sunlight to stored energy?

**2** Where is energy stored?

**3** In what process is stored energy released?

# Producers and consumers

Green plants and algae are green because they contain **chlorophyll**. During photosynthesis chlorophyll absorbs sunlight. The energy is used to make glucose. Further chemical reactions in the cells of organisms make other organic compounds. Organisms that can make glucose and other organic compounds are **producers**.

Carbon dioxide is found in the air. It is also found dissolved in water (oceans, rivers and lakes). Producers use carbon dioxide and water from their environment to make the **sugar**, glucose. This glucose can then be made into other carbohydrates such as starch, or into fats and proteins. These make up the bodies of the producers, forming biomass.

Animals lack chlorophyll. Non-green plants and microorganisms, such as fungi and most bacteria, also lack chlorophyll. So none of these can photosynthesise and make organic compounds. They must feed on the producers or on each other. Organisms that have to obtain organic compounds from the biomass of other organisms are **consumers**.

Carbohydrates, fats and proteins require a source of energy to be made and release energy when they are broken down.

> Only producers can make the organic molecules needed by all organisms.

> All organisms respire – they break down biomass to release energy.

## Remember

Photosynthesis is used to make all the organic molecules found in the bodies of living things.

**FIGURE 3:** Consumers cannot make their own organic compounds. They must feed on other organisms to obtain them.

## QUESTIONS

**4** Why are consumers unable to make organic molecules?

**5** Which organic molecules form biomass?

# Respiration and photosynthesis

Photosynthesis is sometimes summarised by the **chemical equation**:

$6CO_2 + 6H_2O \rightarrow C_6H_{12}O_6 + 6O_2$

Aerobic respiration is sometimes summarised by the chemical equation:

$C_6H_{12}O_6 + 6O_2 \rightarrow 6CO_2 + 6H_2O$

These equations can be misleading as both processes are more complicated. However, the equations do show the overall changes.

The original atmosphere of Earth contained no oxygen. Eventually photosynthetic organisms released enough oxygen into the air to make aerobic respiration possible. Energy was released more efficiently for life processes and organisms began to evolve into more complex forms.

## QUESTIONS

**6** Describe how photosynthesis and aerobic respiration act as opposites.

**7** What happens to energy in each of the processes described in question 6?

# Preparing for assessment: Analysing data and evaluating the investigatio

*To achieve a good grade in science, you not only have to know and understand scientific ideas, but you need to be able to apply them to other situations and investigations.*

Connections: This task relates to Unit 1 context 3.3.2.1 Life on our planet. Working through this task will help you to prepare for the practical investigation that you need to carry out for your Controlled Assessment.

## Investigating effect of light on leaf growth

Sharmila's family had an orchard on a south-facing hill. The sides of the trees facing down the slope received more sunlight than the sides facing up the slope. She had picked some leaves and thought that those from the south-facing side felt heavier, though looked smaller, than leaves from the north-facing side. She wanted to investigate whether the growth of leaves depends on how much sunlight they get.

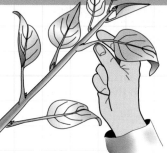

**FIGURE 1**: Sharmila always picked the fourth leaf back from the end of the twig. This was to make sure all the leaves sampled were about the same age.

Sharmila knew that:

> trees need sunlight for photosynthesis

> transpiration is caused by evaporation of moisture from the surface of leaves.

Sharmila picked 10 leaves from the south-facing side of one of the trees and 10 leaves from its north-facing. She measured the surface area of each leaf by putting it on paper divided up into squares with sides of 1 cm and drew around it. She counted the number of full squares inside the leaf outline and the number of part squares inside the leaf outline. She added together the number of full squares and half the number of part squares to give its area.

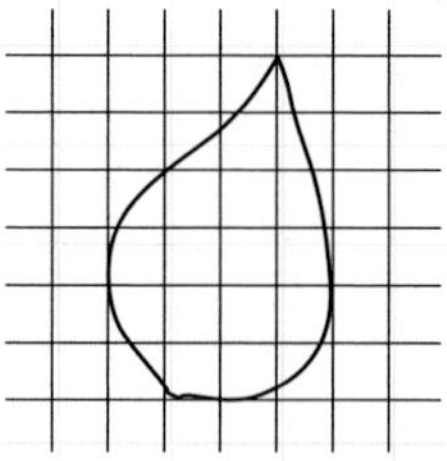

**FIGURE 2**: This is how Sharmila calculated the surface area of a leaf.

To measure the mean thickness of the south-facing leaves, Sharmila stacked them together, measured the total thickness and divided by the number of leaves (10). She did the same thing with the north-facing leaves.

> mean thickness of south-facing leaves = 0.23 mm

> mean thickness of north-facing leaves = 0.18 mm

Sharmila knew the owner of an orchard, also on a south-facing hill, in a neighbouring village. He had carried out a similar set of measurements on a tree in his orchard. Sharmila thought that it would be useful to compare her data with this secondary data.

> mean thickness of south-facing leaves = 0.22 mm

> mean thickness of north-facing leaves = 0.17 mm

| | TREE 1 (primary data) | | TREE 2 (secondary data) | |
|---|---|---|---|---|
| | South-facing trees (most sun) | North-facing trees (most shade) | South-facing trees (most sun) | North-facing trees (most shade) |
| Leaf | Area ($cm^2$) | | | |
| 1 | 14.0 | 29.0 | 12.0 | 29.5 |
| 2 | 18.0 | 19.5 | 16.5 | 22.5 |
| 3 | 13.5 | 28.0 | 17.0 | 26.0 |
| 4 | 21.5 | 29.5 | 18.0 | 33.0 |
| 5 | 15.5 | 23.0 | 14.5 | 22.5 |
| 6 | 12.0 | 32.5 | 20.5 | 29.5 |
| 7 | 16.0 | 21.5 | 18.0 | 20.0 |
| 8 | 20.0 | 20.0 | 13.5 | 27.0 |
| 9 | 19.5 | 22.5 | 12.5 | 19.5 |
| 10 | 16.0 | 24.0 | 17.0 | 24.0 |
| mean | **16.6** | **25.0** | **16.0** | **25.4** |

**TABLE 1**: A summary of the data Sharmila collected.

## ☀ Analysing data

**1.** Sharmila used her own results and the data obtained by her friend in the neighbouring village. Suggest a reason for doing this.

**2.** Use both sets of results to explain Sharmila's first impression – that leaves from the south-facing side of the trees felt heavier, but looked smaller, than ones from the north-facing side.

**3.** What conclusions can you draw about the growth of leaves on south-facing trees compared with those on north-facing trees?

**4.** Compare and comment on the range of measurements made of the area of south-facing leaves and the area of north-facing leaves.

> Make sure that you use the terms *primary data* and *secondary data* in your answer.

> You need to remember that volume is surface area multiplied by thickness and that density is mass divided by volume.

> Sharmila noticed a pattern in the results. Leaves on the south-facing trees had a smaller surface area but were thicker than leaves from the north-facing side. What is different about the conditions in which leaves on the south-facing trees grow compared with the north-facing side? Make sure your conclusions are well structured, clear and logical.

> Remember, *range* is the lowest value up to the highest one.

## ☀ Evaluating the investigation

**5.** Explain whether or not you think Sharmila controlled other variables sufficiently to investigate the difference between leaves growing on south-facing branches and leaves growing on north-facing branches?

**6.** Discuss Sharmila's technique for finding the areas of the leaves and how it might be improved.

**7.** Suggest how Sharmila could improve the technique she used to find the mean thickness of the leaves.

> What were the other variables? You might consider things such as the height of trees and how much they are shaded by other trees. From what height branches should the leaves be picked? Access to water is also a consideration. You may be able to think of others.

> How could Sharmila have improved the resolution of her measuring technique? Remember, *resolution* means the smallest change that the technique can measure.

> Why was measuring the thickness of a pile of leaves easier than trying to measure just one leaf? What might Sharmila have done to make her measurements on the two sets of leaves comparable?

# Food chains and energy flow

**THE SCIENCE IN CONTEXT** Ecologists and other scientists are interested in how organisms in an environment depend on one another. They describe how organisms transfer energy in an ecosystem by using food chains. These food chains are vital to the sustainability of an ecosystem. Organisms in an ecosystem may be categorised as producers or consumers.

## Zebra feast

When lions have feasted on a recently killed zebra, all that remains are a few bones, some scraps of meat and its horns. The vultures are waiting their turn. Where do all of these animals obtain their energy? It all came from sunlight originally.

**FIGURE 1**: The zebra ate grass. Energy from grass is now being transferred to the lion.

## Energy flow

Food chains show how energy passes from one organism to another as they eat and are eaten. The energy is stored in the food that the animals eat. The arrows show the direction of flow of energy and matter from one organism to the next.

```
grass  →  zebra  →  lion
```

**Did you know?**

In one year, one square metre of grass captures about 15 000 kJ of energy from the Sun.

The chemical compounds that store the energy and flow down a **food chain** are carbohydrates, fats and proteins.

When energy is needed, these compounds are broken down by chemical reactions. The products cannot store as much energy as the reactants. The excess energy is transferred to do work.

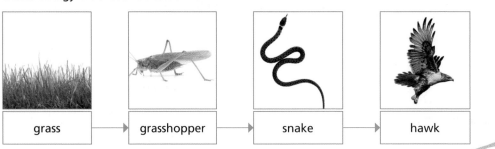

```
grass  →  grasshopper  →  snake  →  hawk
```

**FIGURE 2**: Food chains show the flow of matter and energy between organisms in an ecosystem.

**Remember**

Grass is like a huge power station. It stores energy carried by sunlight. When animals eat the grass, the stored energy is transferred to the animals.

## QUESTIONS

**1** What is the initial source of energy for all the living organisms in a food chain?

**2** What do the arrows in a food chain indicate?

**3** Draw a food chain to show how you get energy from milk.

# Producers and consumers in food chains

Grass is the starting point for numerous food chains. It is a **producer**. It absorbs sunlight and uses the energy to produce carbohydrates and other nutrients. All food chains begin with producers.

The zebra and the lion are **consumers**. They consume food that has been made originally by plants. Chemical compounds in the food eaten by the lion have first of all been consumed by the zebra.

The zebra is a **primary consumer**, because it is the first consumer in the food chain. It is also a **herbivore**. Herbivores eat plants.

The lion is a **secondary consumer**, because it is the second consumer in a food chain and consumes a primary consumer. It is also a **carnivore**. Carnivores eat other animals.

The lion is also a predator. Predators kill and eat other animals for food. The animals that they kill are their **prey**.

## QUESTIONS

**4** In a freshwater pond, tadpoles eat algae and diving beetles eat tadpoles. In the food chain, which is (a) the producer (b) the primary consumer (c) the tertiary consumer?

**5** Draw a food chain to show how the diving beetles get energy.

## Life in water

In water the producers are **algae**. They possess chlorophyll just like green land plants. They vary in size from minute single-celled organisms to large seaweeds.

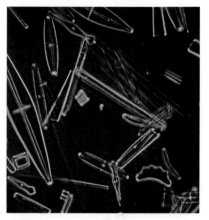

FIGURE 3: Microscopic algae can carry out photosynthesis, just like grass and other plants.

# Energy wastage

Figure 4 shows sunlight filtering through the leaves of a tree. The photographer was standing on the ground looking upwards.

The sky can be seen through the leaves. Some of the light is missing the leaves altogether, going straight down to the ground beneath the tree. You can see through some of the leaves, so some light must be going right through them and reaching your eyes.

Green plants capture only a small amount of the energy from sunlight that falls onto them. This is because some light:

> misses the leaves altogether

> reaches the leaf and is reflected from the leaf surface

> reaches the leaf but goes all the way through without hitting any chlorophyll

> reaches the chlorophyll, but is not absorbed because it is of the wrong wavelength (colour).

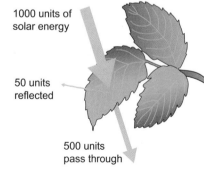

FIGURE 4: Do the leaves of a tree catch all of the sunlight?

1000 units of solar energy

50 units reflected

500 units pass through

FIGURE 5: This is what happens to the energy from sunlight that falls onto a leaf.

## QUESTIONS

**6** Look at Figure 5. Out of 1000 units of energy carried by sunlight, how much can the plant use for photosynthesis?

**7** What colour of light does chlorophyll reflect without absorbing it?

# Pyramids of biomass

**Essential notes**

> food chains show the flow of matter and energy between producers and consumers

> all of the energy used by living organisms comes from the Sun

**THE SCIENCE IN CONTEXT** All of the living organic matter produced in a given area is called the biomass. As well as being interested in energy flow, ecologists investigate what happens to biomass as it passes along the food chain. They monitor the numbers and sizes of the organisms in food chains and construct a scale drawing called a pyramid of biomass.

## Gentle giant

Whale sharks are the largest living species of fish. They can be up to 20 metres long. Despite their size, whale sharks feed on plankton, the microscopic organisms in water.

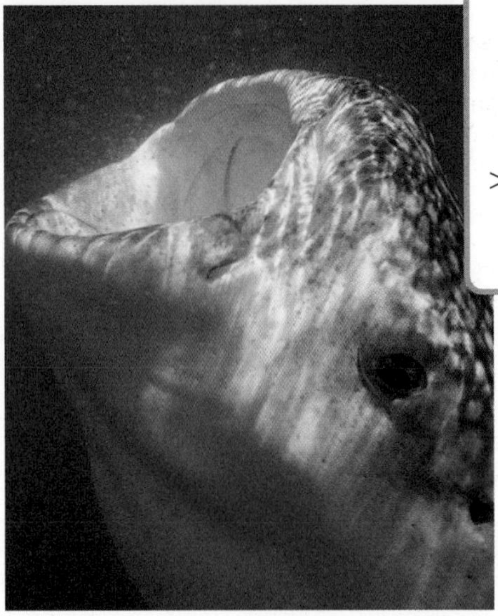

**FIGURE 1**: Why does a whale shark have such an enormous mouth?

## Biomass and food chains

Think about this food chain:

grass → antelope → lion

How many antelope are needed to support one lion? It probably needs several hundred in its lifetime.

How much grass is needed to support one antelope? It will be much more than the antelope's own body mass.

If the mass of:

> all the grass needed to supply the antelope, and

> all the antelopes needed to support the lion

was measured, something like Figure 2 might be obtained.

The mass of living material is called biomass. Figure 2 shows a **pyramid of biomass**.

> The width of each block represents the **biomass** at each step in a food chain.

> At each stage there is less biomass than in the stage before.

> This is often about only one-tenth of the previous energy or biomass.

Why does the biomass reduce?

The antelope do not eat all the grass. For example, the roots of the grass are under the ground, so not all the grass biomass is eaten by the antelope.

Secondly, not all the antelope are killed and eaten by lions. Even if the lion does kill an antelope, it does not eat it all. So, not all of the antelope biomass is passed on to lions.

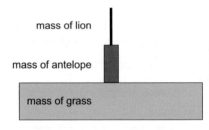

mass of lion

mass of antelope

mass of grass

**FIGURE 2**: A pyramid of biomass.

### Did you know?

Each hour 6000 litres of seawater are filtered through a whale shark's gills.

### QUESTIONS

**1** Suggest two reasons why not all the grass biomass is passed on to antelopes.

**2** Which parts of an antelope are not eaten by lions?

**3** Draw a pyramid of biomass that has a whale shark at the top.

# Mass and energy changes

Measuring biomass as the mass of living material can be difficult. This is because a high proportion of living **tissue** is water and can vary quite a lot. Biomass is usually considered to be the mass of the organic compounds in organisms – this is the total living organic matter.

It can be estimated by drying samples, so that the water is not included in the figure obtained. This gives a good comparison of the biomass that is passed on in a food chain.

Whenever energy is transferred, some is **wasted**. The transfer is never 100% **efficient**. At each step of a food chain, some energy being transferred heats up the surroundings. The **'wasted' energy** is not used by the living organisms.

At each stage along a food chain there is less biomass available for the organisms to use and, therefore, less of the stored energy is available to pass along.

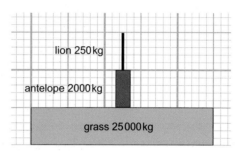

lion 250 kg

antelope 2000 kg

grass 25 000 kg

**FIGURE 3**: A pyramid of biomass drawn to scale.

## QUESTIONS

**4** Explain why biomass and the amount of energy are less at each stage of a food chain than they were at the previous stage.

**5** The biomass of algae in a pond is 1000 g. The biomass of tadpoles is 100 g. The biomass of diving beetles is 10 g. Draw a pyramid of biomass for this food chain. You will need to use graph paper, and work out the scale to use.

# Food webs

A food chain shows just one pathway that energy takes as it passes from one organism to another. In reality one type of organism is often fed on by more than one other consumer.

You can add other organisms to food chains and start to build up a food web. For example, jackals could be added to the grass → antelope → lion food chain.

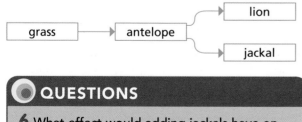

grass → antelope → lion

antelope → jackal

**FIGURE 4**: A jackal waits to see what the lion leaves behind.

## QUESTIONS

**6** What effect would adding jackals have on a pyramid of biomass?

# Efficiency and decay

**THE SCIENCE IN CONTEXT**
All organisms eventually die and decay. The decay process recycles useful materials such as plant nutrients. Microorganisms are very important in many decay processes. The recycling is put to good use by food growers such as farmers and market gardeners. They put rotted waste on their land to replace nutrients used when the previous crop grew.

### Bog body

The body in the photo was found in a peat bog. It is a man who died during Roman times about 2000 years ago. His stomach contained the remains of his last meal – some bread.

FIGURE 1: Lindow man from a peat bog in Cheshire.

## Decay and recycling

**Decay** happens when organisms feed on a dead body or waste material from animals and plants. These organisms are decomposers. Many decomposers are microscopic bacteria and fungi that produce enzymes. The enzymes they produce digest the dead material.

Water in peat bogs prevents microorganisms from getting enough oxygen for respiration. Once a body is buried in peat, it decays only very slowly.

### Recycling

Decay by decomposers is a vital process.

> If things did not decay, we would be knee-deep in dead plants, animals and waste.

> The decay process releases minerals into the soil which plants need to grow. There is a balance between the processes that remove materials from the environment, and the processes that return materials to the environment.

FIGURE 2: The blue mould is breaking down the lemon using enzymes.

### Did you know?

Herbivores usually obtain less energy from feeding on plants than decomposers obtain from decaying their dead remains.

### QUESTIONS

1 What is a decomposer?

2 How can decaying waste help plant growth?

3 Explain why the body in the bog did not decay.

## Recycling and food chains

Figure 3 shows how decay microorganisms fit into a simple food web and feed on every organism in the chain. They break down the waste material that plants and animals produce, and then finally their dead bodies.

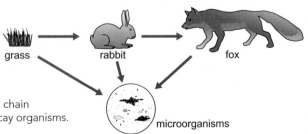

FIGURE 3: A food chain including the decay organisms.

grass    rabbit    fox    microorganisms

Q ... gcse decay

## Compost heaps

Gardeners often make **compost** heaps of garden or kitchen waste. Microorganisms digest the materials in the plants to form a crumbly brown substance. This compost can be put onto the garden, to provide minerals that help plants to grow.

## Speeding or slowing decay

Anything that affects microorganisms can affect the rate of decay.

> Temperature: Microorganisms function better at warm temperatures because enzymes catalyse their metabolic reactions, just like ours. Decay happens fastest at the optimum temperature for enzymes, usually between 25 °C and 45 °C.

> Moisture: Microorganisms need moisture in order to reproduce and feed. Some kinds can survive when it is very dry, but they will not be able to reproduce until water is available.

> Oxygen: Many microorganisms need oxygen for respiration. They are more active when oxygen is plentiful.

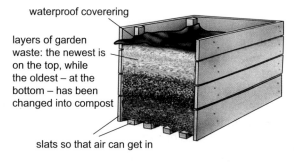

waterproof coverering

layers of garden waste: the newest is on the top, while the oldest – at the bottom – has been changed into compost

slats so that air can get in

**FIGURE 4**: Suggest how the slats help the microorganisms to make the compost.

### QUESTIONS

**4** Explain why a good compost heap is warm and moist.

**5** Suggest how having a compost heap can help the environment.

## Efficiency

Eventually all energy leaves the biosphere. It is transferred to the surroundings which get warmer. Whenever energy is transferred, some can no longer be used by organisms.

Figure 5 shows how energy transfer through a food chain is inefficient because energy used in respiration is wasted through heating. Other energy is wasted in plant material that passes out of the bodies of herbivores in faeces. This energy and the energy in dead remains pass to decomposers.

To calculate the **efficiency** of energy transfer:

$$\text{efficiency} = \frac{\text{useful energy transferred}}{\text{original amount of energy}} \times 100\%$$

For example, a leaf uses energy in photosynthesis, to make carbohydrates and other molecules. If 200 units of energy from sunlight hit the leaf, but it uses only 40 units:

$$\text{efficiency} = \frac{40}{200} \times 100 = 20\%$$

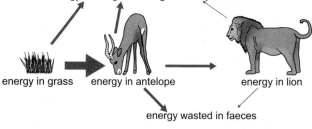

energy wasting in heating from respiration

energy in grass    energy in antelope    energy in lion

energy wasted in faeces

### QUESTIONS

**6** Calculate the efficiency of energy transfer in this leaf.

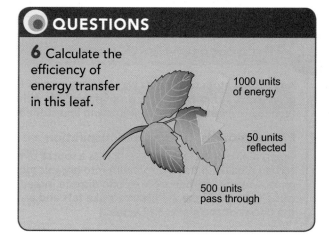

1000 units of energy

50 units reflected

500 units pass through

*To achieve a good grade in science, you not only have to know and understand scientific ideas, but you need to be able to apply them to other situations and investigations. These tasks will support you in developing these skills.*

## ✷ Wandering wildebeest

Ashley was watching a wildlife programme on TV. It was about the Serengeti National Park in Africa and featured animals called wildebeest. He was fascinated by the odd appearance of these animals, with their large heads, curved horns, big shoulders and spindly rear legs; then the programme explained that their journey is the biggest mammal migration in the world.

One reason for the migration is the need for fresh grazing areas – short grass is their preferred diet. The wildebeest cannot go without water for more than a day or so. They have several predators, including hyena and lions. An adult will live for up to 20 years.

The females calve in May, giving birth to a single calf; they don't seek shelter but give birth surrounded by the herd. Most of the females in the herd give birth within two or three weeks of each other. The calf can stand and run within minutes of being born. Within a few days, it can run fast enough to keep up with the herd.

The herd starts to migrate north soon after that and travels at a relentless pace through day and night; many are lost, injured or even killed. In November the return journey south commences.

Ashley was intrigued by the programme and could not understand why the wildebeest lived the way that they did. The young seemed to have a really rough time; they had to be up and on the move in a very short time and sometimes perished before reaching adulthood.

Surely they would stand a better chance of survival if they were not all born at about the same time or, if the herd stayed in the same area for several months, until the calves were older and stronger?

It made a really good programme, but he was so glad that humans didn't raise their young like that.

## ✳ TASK 1

What is the preferred diet of wildebeest and why must they live close to water?

Why do wildebeest have to move on repeatedly?

## ✳ TASK 2

If the wildebeest lived in smaller herds all the year round (instead of just during the breeding season), thought Ashley, perhaps they would be able to settle in one area. Why do you think they do not do that?

## ✳ TASK 3

(a) Why do you think the females give birth in the middle of the grassy plain, surrounded by the herd, instead of finding a sheltered place?

(b) Most of the females give birth at around the same time. Why does this give an advantage, in terms of survival?

## ✳ TASK 4

Ashley found the mass migration a stunning sight, with a great number of animals relentlessly pressing on. They did not stop, even if some members of the herd were injured or left behind. Surely they would stand a better chance of survival, if they cared more for each other? What do you think?

## ✳ TASK 5

Use some of the ideas that you have gathered, to explain how the need for survival has influenced the adaptation of the wildebeest.

## ✳ MAXIMISE YOUR GRADE

| | Answer includes showing that you can... |
|---|---|
| **E** | give one reason why the wildebeest keep moving on. |
| | give one reason why large group size aids survival. |
| | give several reasons why the wildebeest keep moving on. |
| | suggest several reasons why large group size aids survival. |
| **C** | explain why being out in the open aids survival. |
| | explain why many females giving birth at the same time aids survival. |
| | explain, with reference to one feature of wildebeest behaviour, how survival has influenced them. |
| **A** | explain, with reference to both examples of wildebeest behaviour and features, how survival has influenced them. |

73

## To achieve your forecast grade in the exam you will need to revise

Use this checklist to see what you can do now. Refer back to the relevant pages in this book if you are not sure. Look across the three columns to see how you can progress.

Remember that you will need to be able to use these ideas in various ways, such as:

> interpreting pictures, diagrams and graphs

> applying ideas to new situations

> explaining ethical implications

> suggesting some benefits and risks to society

> drawing conclusions from evidence that you are given.

Look at pages 250–271 for more information about exams and how you will be assessed.

| To aim for a grade E | To aim for a grade C | To aim for a grade A |
|---|---|---|
| Know that life can be categorised into kingdoms. | Understand that animals and plants can be classified according to their physical characteristics. | Explain why classification is important as an international method of grouping living organisms to help name and identify them. |
| Know that organisms obtain the materials that they need to survive from their surroundings and from other living organisms. Know that extremophilic microbes have been found living in the Arctic, volcanic vents, very dry environments and severe chemical environments. | Explain how animals, plants and microbes are adapted for survival in the conditions in which they normally live. | Explain how plants can adapt to conditions through changes in surface area, water storage tissues and extensive root system. Explain how animals can adapt using surface area, insulation, body fat and water storage systems. |
| Know that evolution occurs through natural selection. | Explain that individuals with characteristics most suited to the environment are more likely to survive and breed and less well adapted individuals are more likely to die without breeding. | Explain how the genes that have enabled better-adapted individuals to survive are passed on to the next generation more often than genes that make organisms less well adapted. |

## To aim for a grade E    To aim for a grade C    To aim for a grade A

Describe how light, temperature, day length and gravity affect plant growth.

Know that auxins control plant growth.

Explain how phototropism and gravitropism are controlled by auxins.

---

Know that energy enters the biosphere as sunlight.

Know that in photosynthesis, energy from sunlight is stored in organic compounds (biomass) by producers.

Know that biomass is broken down to release energy through respiration by consumers.

Know that energy leaves the biosphere by heating the surroundings.

---

Understand that food chains show the flow of matter and energy between the producers and consumers in an ecosystem.

Know that the mass of living material (biomass) and amount of energy at each stage in a food chain is less than it was at the previous stage.

Understand that when organisms die, decomposers feed on their remains.

Calculate the percentage of energy transfer at each stage of a food chain.

Explain the reasons for the inefficiency of the energy transfer in a food chain.

---

Know that carbon dioxide is removed from the environment in photosynthesis.

Know that the carbon from the carbon dioxide is used to make carbohydrates, fats and proteins.

Know that when producers are fed on by animals or decomposers, some of the carbon becomes part of the fats and proteins that make up their bodies.

Understand that when living organisms respire, some of their carbon becomes carbon dioxide and is released into the atmosphere.

Know that carbon is stored in fossil fuels and is released as carbon dioxide when they are burned.

---

Know that when living things die, their bodies are broken down by decomposers, releasing the elements they contain, which are then used by plants to grow.

Know that microorganisms function better in warm, moist conditions and in a plentiful supply of oxygen.

Use data to construct pyramids of biomass to scale.

---

Describe how limestone (calcium carbonate) is formed from carbon dioxide dissolved in water.

Know that carbon deposited on the sea floor is removed from the rest of the carbon cycle for a long period of time.

## ✳ WORKED EXAMPLE – Foundation tier

**1.** A mineral can be an element (metal or non-metal) or a compound which is found naturally in the Earth's crust. Gold, sulfur and haematite are examples of commonly used minerals. Haematite is a black mineral containing mainly iron oxide. A metallic ore is a mixture of a metallic mineral mixed with other materials from surrounding rocks.

**(a)** Using the information above, classify gold, sulfur, iron oxide and haematite, as elements mixtures or compounds, giving a reason for your choice. [4]

*Gold is a metal, sulfur is a non-metallic element and iron oxide is a compound because they are found in the earth's crust. Haematite is a metallic ore.*

**(b)** Haematite is mined worldwide and the process needs to be economical. State two factors to be considered when mining an ore. [2]

*It's got to have enough of the metal, or one of its compounds, in it to be worth digging out and there needs to be enough of the ore to set up the mine in the first place.*

**(c)** Many metals are found combined with non-metals like oxygen (oxide ion) and reduction is needed to obtain the metal from the oxide. Explain what happens in reduction. [2]

*Reduction is when the ore becomes less*

**(d)** Name a reducing agent that can be used to extract iron from iron oxide. [1]

*Carbon*

**(e)** The reactivity series can be used to decide the best method to extract the metal from its oxide. Here is part of the series, going from reactive to less reactive: **sodium, magnesium, aluminium, carbon, zinc, iron, tin, lead.**

Using the series, suggest a method for extracting aluminium and magnesium from their ores. Explain your answer. [2]

*Can be extracted using an electric current. The method is chosen because of the position in the reactivity series.*

### How to raise your grade!
Take note of these comments – they will help you to raise your grade.

The candidate has classified gold as a metal but needs to say that it is an element. Iron oxide and sulfur have been correctly classified, but the reason is incorrect. The candidate should explain that an element is made from just one type of atom, but a compound is made from two or more different types of atoms. Haematite is a mixture as it is made of mainly iron oxide mixed with other rocks. The candidate will obtain two marks.

The candidate has answered this question correctly as he has understood the economic factors involved in mining an ore. The candidate will obtain both marks.

This is not a good answer. The correct answer is that reduction is the removal of the oxygen from the metal oxide to give the metal.

Answer is correct. Carbon monoxide would also have been correct.

This answer does not quite answer the question – but may receive one mark. Aluminium and magnesium are extracted using an electric current, but the actual method of electrolysis is needed. The candidate should have explained that metals more reactive than carbon need to be extracted using electrolysis.

## ✪ WORKED EXAMPLE – Higher tier

*In this question you will be assessed on using good English, organising information clearly and using specialist terms where appropriate.*

**Phytomining is a way to extract metals and to generate revenue.**

Certain plants can extract significant quantities of metals from soil. One team of scientists is using alyssum plants to extract nickel and cobalt. Plants are harvested, burned and the metals recovered from the ash. Research is continuing.

It is estimated that metal-contaminated land might yield only about £30 to £100 per hectare annually. However, phytomining plants could yield about 400 kilograms of nickel, enough to produce an income of over £1500 (increasing to over £2000 if energy produced by the biomass is sold).

In the future, phytomining might be used where traditional methods would be uneconomic.

For a presentation on metal extraction, make three slides to compare phytomining with traditional methods of metal extraction. [6]

> *Slide 1: Extraction of metals from their ores depends on the reactivity of the metal. The more reactive it is the harder it is to extract. Metals like Aluminum need to be extracted by using electricity using electrolysis. This is expensive and uses a lot of energy.*

> *Slide 2: Less reactive metals like lead or iron can be extracted using carbon. These can be extracted when the oxide turns into the metal. This process needs a lot of heat and produces carbon dioxide.*

> *Slide 3: Phytomining is the process which can be used to extract metals from the soil. Plants are grown on mining waste. The plants are harvested, dried and burned. The metal is extracted from the ash.*

### How to raise your grade!
Take note of these comments – they will help you to raise your grade.

The information on the slides is clear. There is some good science although the candidate has not mentioned extraction from the ore or reduction.

There could be more scientific detail on carbon as a reducing agent and possibly an equation.

There is good extraction of information from the text – but some comparison would enhance the answer.

This answer would gain five marks.

## What you should know

### Nerves and hormones

The nervous system contains sense organs, which use nerves to send signals to the brain and spinal cord.

The brain and spinal cord send signals to muscles to control what they do.

The endocrine system contains glands, which release hormones into the blood to control many body systems.

 Name your five senses.

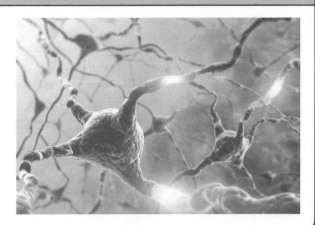

### Food

The three main food groups are carbohydrates, proteins and fats.

Enzymes are used to break large food molecules into smaller molecules which can be absorbed.

Into which small molecules are carbohydrates, proteins and fats broken down?

### Genetics

Variation in individuals may be due to genetic or environmental causes or a combination of both.

Genes may be dominant or recessive.

Describe one of your characteristics that is not affected at all by the environment.

# You will find out

## Control of body systems

> Receptors respond to particular types of stimuli to trigger electrical impulses in neurones.

> Reflex actions use sensory, relay and motor neurones to generate rapid, automatic responses.

> Insulin and glucagon are used to regulate blood sugar levels.

> In diabetes, blood sugar levels rise too high, but this can be controlled by insulin injections for Type 1 and changes in lifestyle for Type 2.

> Body temperature is controlled by the thermoregulatory centre in the brain, which alters the rate of heat loss from the body.

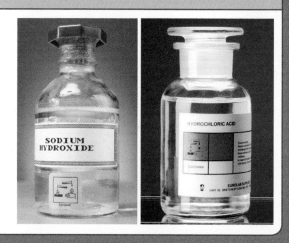

## Chemistry in action in the body

> Acids and bases are hazards, but the risks of using them can be minimised.

> In neutralisation reactions, an acid and an alkaline substance react to give a salt plus water.

> Excess stomach acid can be neutralised by antacids.

## Human inheritance and genetic disorders

> Chromosomes contain genes, which control body characteristics.

> Genes have different forms called alleles, which produce different characteristics.

> Monohybrid inheritance involves dominant and recessive alleles.

> Cystic fibrosis, sickle cell anaemia and haemophilia are inherited disorders caused by faulty alleles.

# Receptors

**THE SCIENCE IN CONTEXT** Senses (sight, hearing, smell, taste and touch) help to keep you alive and enjoy life. How sensitive a person is to these stimuli may vary from one sense to another. For example, somebody might have poor eyesight, but a strong sense of smell. How well your senses work can give healthcare workers information about your health.

## Distinguishing smells

Our noses can distinguish 10 000 odours. Yet it took a team of scientists ten years to find molecules which could distinguish just fifty of them.

### Essential notes

> a receptor responds to a particular type of stimulus and triggers electrical impulses in neurones

> information is sent to the brain, which coordinates the body's responses

**FIGURE 1**: Why are scientists developing 'artificial noses'?

## Receptors

Your senses help you to stay alive. They also help you enjoy life.

A **stimulus** is a change in surroundings. **Receptors** are specialised cells in your body. These cells detect stimuli. They are often found in sense organs.

Each type of receptor is sensitive to a particular kind of stimulus. If a stimulus is detected, the receptor triggers electrical impulses in a **neurone** (nerve cell). Information is sent to the brain, which coordinates the body's responses.

### The five senses

Humans have five 'senses'. Each sense uses different kinds of receptors. They detect many stimuli.

Doctors and other health workers check your senses. This helps them to assess your health.

### Did you know?

Parmesan cheese and vomit share a distinctive smell because both contain butyric acid.

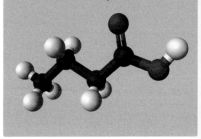

| Sense | Organ | Stimulus detected | Type of receptor | What receptor does |
|-------|-------|-------------------|------------------|---------------------|
| sight | eye | light | photoreceptors in retina | absorbs light energy |
| hearing | ear | sound | sound receptors in cochlea | detects vibrations in fluid |
| smell | nose | chemicals in air | smell receptors in nasal cavity | detects the shape of molecules |
| taste | tongue | chemicals in food | smell receptors in taste buds | detects the shape of molecules |
| touch | skin | pressure, temperature and pain | different kinds of nerve ends | detects pressure, hot, cold or pain |

## QUESTIONS

**1** How do you become aware of a change in your surroundings?

**2** For each sense, suggest one way in which it can help to protect us from harm.

**3** Which receptors detect chemical stimuli?

Q ... gcse receptors and effectors

# Light, sound, smell, taste and touch

### Detecting light

Receptors in the retina are **rods** and **cones**.

Rods and cones contain pigments. The pigments **absorb** visible light and lose their colour. This triggers nerve impulses that travel through the optic nerves to the brain.

Cones are sensitive to colour. They work best in bright light. There are three kinds, which absorb mainly red, green or blue **wavelengths**. Other colours are distinguished by how much light each cone absorbs.

Rods detect shades of grey only. They become more sensitive in dim light. You can see the shapes of objects in a dimly lit room, but the colours appear washed out.

### Detecting sound

The **ear** detects sound from vibrations in the air. It changes these to vibrations in fluid in the **cochlea**. These are detected by hearing receptors.

### Detecting smell

Chemicals, in the air, dissolve in the thin layer of mucus that covers smell receptors in the nasal cavity. About 1000 different kinds of receptor combine to distinguish 10 000 different odours. The receptors are very sensitive and can detect just a few **molecules**.

### Detecting taste

Some people are 'supertasters' (35% of women and 15% of men). They have extra taste buds. Bitter tastes, such as espresso coffee or olives, are too strong for them.

The tongue has about 1000 taste buds, each containing about 100 receptor cells. They distinguish salty, sour, sweet or bitter tasting molecules. Savoury has recently been added to the list. The heat of chilli is caused by stimulation of pain receptors in the tongue!

Food would seem very bland if you only had taste receptors. The 'taste' of food is actually a combination of taste and smell.

### Detecting touch

Specialised touch receptors in the skin detect pressure, pain, hot or cold. Most are bare nerve endings which detect pain. Others include Meissner's corpuscles – very sensitive capsules found in 'ticklish' areas.

FIGURE 2: Suggest why the sense of smell in humans is less sensitive than in a dog.

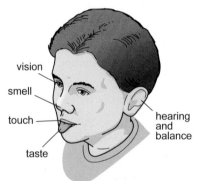

vision

smell

touch

hearing and balance

taste

FIGURE 3: Why are so many of your sense organs located in your head?

## QUESTIONS

**4** Describe a supertaster.

**5** Why are fingertips very touch sensitive?

# Transducers

A transducer is a device which changes one form of input into a different form of output. This involves the transfer of **energy**. For example, a loudspeaker transfers energy from electrical **current** into sound.

A stimulus causes a change in a receptor and triggers a nervous impulse in a neurone. This is an electrical signal, which is generated using energy from respiration in the neurone.

## QUESTIONS

**6** Explain why receptors are not transducers in the usual sense.

# Reflex action

**THE SCIENCE IN CONTEXT** In a healthy body, sense organs detect external changes. The brain processes messages sent to it. The brain then sends a response that causes the body to make physiological changes. Doctors find out about a person's health by, for example, checking knee jerk reaction to a sharp tap or an eye's pupil reaction to bright light.

## Robot control

For a robot to move around obstacles or climb stairs, and mimic human reflex actions, it needs receptors (detecting changes in the surroundings) and effectors (making the robot respond).

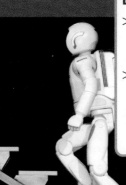

**FIGURE 1**: Asimo, a robot, has artificial intelligence.

### Essential notes

> reflex actions are automatic, rapid responses to stimuli

> reflex actions use sensory, relay and motor neurones to connect receptors to effectors

## The nervous system and coordination

### The nervous system

Your nervous system rapidly controls the reaction of your body to outside stimuli.

The nervous system is made up of **neurones** (nerve cells). These may be:

> bundled into **nerves**

> collected together in the **brain** and **spinal cord**. These form the **central nervous system** (CNS).

Nerves carry fast-moving nervous impulses – electrical signals carrying information.

### Coordination

Receptors detect stimuli.

**Sensory neurones** carry nervous impulses to the brain or spinal cord. This is information about the stimuli – what they are and how strong they are.

The brain or spinal cord coordinates a response by sending nervous impulses through **motor neurones** to **effectors**.

Effectors are muscles or **glands**. They cause something to happen in the body in response to the stimulus.

Often the response is automatic, but sometimes you may think and make choices.

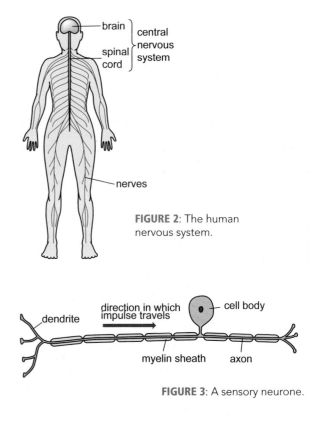

**FIGURE 2**: The human nervous system.

**FIGURE 3**: A sensory neurone.

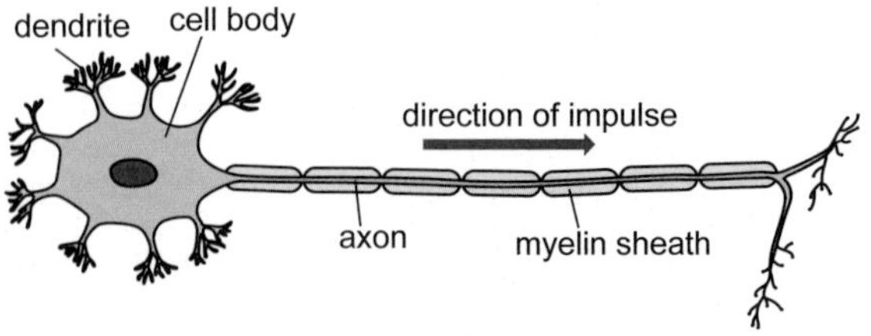

**FIGURE 4**: A motor neurone.

### QUESTIONS

1 Name two parts of the central nervous system.

2 Where do sensory neurones carry nervous impulses from and to?

3 Give the two different kinds of effectors that can carry out responses to stimuli.



Q ... reflex action

# Reflexes

## Reflex actions

Suppose you touch a hot object with your hand. You need to remove your hand as quickly as possible, to prevent getting hurt too badly.

Reflex actions are rapid and automatic – you do not have to learn how to make them. You are born with sets of receptors connected through the nervous system to effectors. The pattern of the **reflex action** is:

1. A receptor detects a stimulus.

2. Electrical impulses are sent through a sensory neurone to the CNS (usually the spinal cord).

3. The CNS sends electrical impulses through a **relay neurone** to a motor neurone.

4. Electrical impulses are sent through the motor neurone to an effector.

5. The effector, a muscle or gland, carries out an automatic response to the stimulus.

Stand on a pin and your leg muscles contract to pull the foot away as quickly as possible. Other reflex actions include blinking, if something touches the front of your eye. This is coordinated by the brain.

## Knee jerk reflex

A doctor may 'test your reflexes' by giving a sharp tap to the tendon below the kneecap. This should cause the leg to kick – the knee jerk reflex. Receptors, in the muscle in the front of the thigh, are stretched and a reflex action causes the muscle to contract. This reflex is unusual as it only uses two neurones, one sensory and one motor.

Stretch receptors in muscles allow the body to maintain posture and in standing and walking – the body automatically controls the muscles.

FIGURE 5: How might a knee jerk reflex show that something is wrong?

## QUESTIONS

**4** What is a reflex action?

**5** Stand upright and start to lean back. Describe how your body will automatically adjust to restore your balance.

# Reflex arcs and synapses

Reflex arcs can coordinate responses to stimuli inside the body, such as:

> constriction of the pupil of the iris, if bright light enters the eye

> sweating, in response to a raised body temperature.

A **reflex arc** is the route taken by a nervous impulse as it travels from a receptor to an effector.

Neurones do not actually join together. There is a very small gap called a **synapse**. Electrical impulses cannot pass across the gap. Instead a chemical is secreted and diffuses across. This stimulates a new electrical impulse in the next neurone.

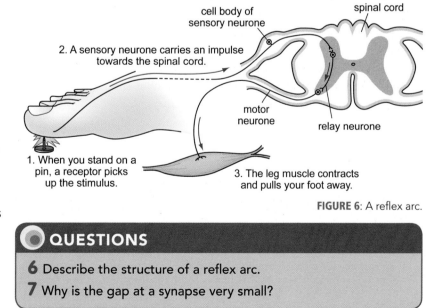

cell body of sensory neurone

spinal cord

2. A sensory neurone carries an impulse towards the spinal cord.

motor neurone

relay neurone

1. When you stand on a pin, a receptor picks up the stimulus.

3. The leg muscle contracts and pulls your foot away.

FIGURE 6: A reflex arc.

## QUESTIONS

**6** Describe the structure of a reflex arc.

**7** Why is the gap at a synapse very small?

# Hearing

**THE SCIENCE IN CONTEXT** Sound is one of the five senses. Ears are designed to detect sound waves. Hearing tests are usually done by audiologists. You might have heard of ear, nose and throat specialists. They are called otolaryngologists and diagnose and treat disorders and diseases of ears, noses and throats.

## Treating deafness

Deafness affects many people, for many reasons. If the ear's hearing receptors still work a little, a hearing aid can be used. It amplifies sounds and transmits them to the cochlea, through the bones in the skull.

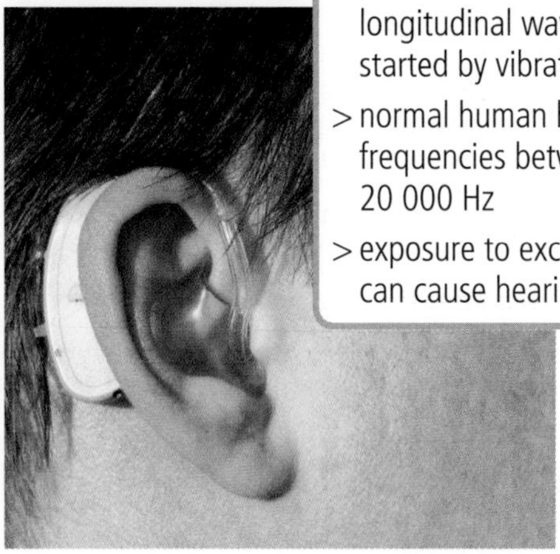

**FIGURE 1**: A hearing aid.

**Essential notes**

> sound travels to the ear as longitudinal waves in the air started by vibrating objects

> normal human hearing detects frequencies between 20 and 20 000 Hz

> exposure to excessive noise can cause hearing impairment

## Making, hearing and describing sound

### Making and hearing sound

Sounds are caused by things vibrating. This could be a tuning fork, a loudspeaker in a music system or your vocal cords when you talk.

Hearing receptors, in the ear, are sensitive to these vibrations.

Vibrations make pockets of air particles vibrate. Sound transfers energy as a wave of vibrating air pockets. You hear sounds when the vibrating air pockets make your eardrum vibrate.

### Sound waves

Sound waves are **longitudinal waves** – air particles vibrate backwards and forwards in the direction that the sound travels.

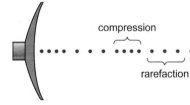

**FIGURE 2**: The vibrating loudspeaker makes the air particles vibrate.

You can picture this by looking at how longitudinal waves pass along a spring. Sections of the spring are pushed together and then pulled apart as a wave travels from one end to the other.

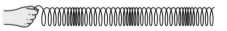

**FIGURE 3**: Longitudinal waves stretch and squash the spring.

### Describing sounds

To describe a sound, you need to know its **amplitude** and **frequency**.

> Amplitude (or volume) – how loud it is. Very loud sounds can damage your hearing.

> Frequency (or pitch) – how high or low it is. Frequency is measured in cycles per second (Hertz, Hz). Humans can hear frequencies in the range 20–20 000 Hz.

**Did you know?**

Sound travels at about 340 metres per second at sea level. That is 3400 times faster than the world's fastest sprinters.

**Remember**

The unit for frequency is the hertz (Hz). One hertz means one cycle per second.

**QUESTIONS**

1 What causes sound?

2 How do sound receptors detect sound?

3 Describe a longitudinal wave.



Q ... gcse sound waves

# The ear, wavelength and frequency

## The ear

The ear collects and amplifies sound waves so that they can be detected. It has three main parts: the outer, middle and inner ear. Sound receptors are in the cochlea, in the inner ear.

Sound wave vibrations:

> are collected by the outer ear

> make the eardrum vibrate

> are amplified in the middle ear

> pass through fluid in the snail-shaped cochlea

> are detected by sound receptors – rows of cells with hairs

> pull on the hairs.

The hairs are attached to a stiff membrane. When the membrane vibrates, electrical impulses are generated in neurones. These nervous impulses:

> contain information about the frequency, intensity (loudness) and duration of the sounds

> pass into the hearing nerve and travel to the brain.

You become aware of the sounds when the electrical signals reach your brain.

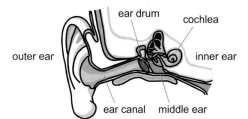

FIGURE 4: Structure of the ear – you do not need to remember this, but it shows you where the action is.

## Wavelength and frequency

A sound wave is characterised by its **wavelength**. In a fixed period of time:

> The shorter the wavelength, the more cycles (number of waves per second). The frequency is higher.

> The longer the wavelength the fewer cycles – the frequency is lower.

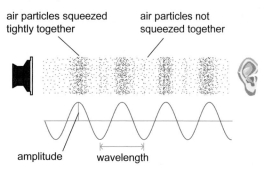

FIGURE 5: Wavelength is the distance between a point on one wave and the same point on the next wave.

Wavelength and frequency are connected by the equation:

$$v = f \times \lambda$$

where

$v$ is the speed of sound – 340 metres per second (m/s)

$f$ is frequency in hertz (Hz)

$\lambda$ is wavelength in metres (m)

Musicians use the term **pitch** instead of frequency. High-pitched sounds have high frequencies and short wavelengths – the waves are spaced closer together. The longer the wavelength becomes, the lower the frequency and the pitch of the sound.

## QUESTIONS

**4** How do sound receptors detect sound?

**5** Explain what 'wavelength' means.

**6** Do high-pitched sounds have waves that are close together or far apart?

# Hearing impairment

Hearing is impaired when an individual can no longer hear sounds normally. Audiologists are healthcare professionals who are trained to identify and treat hearing problems.

## QUESTIONS

**7** Explain the properties of sound that an audiologist could use to test your ability to hear?

Q ... gcse structure of the ear

# Hormones

**THE SCIENCE IN CONTEXT** Glucose is the main source of energy for respiring cells. However, too much glucose in the blood can be unhealthy. A healthy body is able to control the amount of sugar in its bloodstream. It does this by producing a hormone called insulin. Glucose levels that remain high may be caused by diabetes.

### Insulin from bacteria

Scientists can use genetic engineering to transfer genes from one species into another. Since 1978, genetically modified bacteria have been used to make the hormone human insulin, used to treat diabetes.

INSULIN

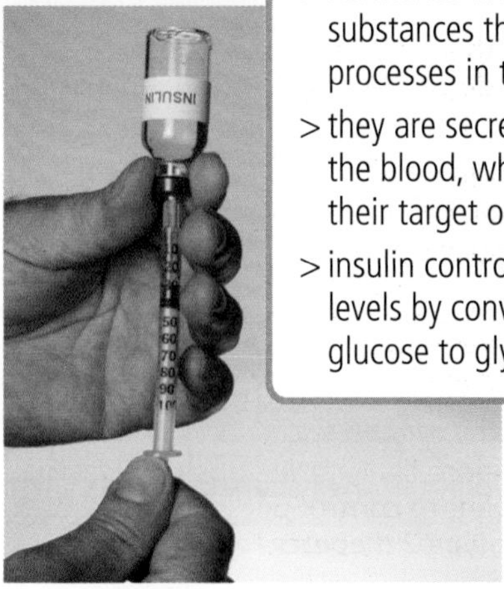

**FIGURE 1**: Why might insulin be needed as a drug?

**Essential notes**

> hormones are chemical substances that control many processes in the body

> they are secreted by glands into the blood, which carries them to their target organs

> insulin controls blood glucose levels by converting excess glucose to glycogen in liver cells

## Hormones and glucose

### Making hormones

**Hormones** are made by **glands** such as the **pancreas**. When a gland releases a substance, it *secretes* it. The pancreas makes and secretes a **secretion** – the hormone **insulin**.

Hormones are chemical substances secreted into the blood. They control many processes in your body. Blood transports hormones all over the body. Each hormone affects only certain **target organs**. The target organ for **insulin** is the liver. This regulates **blood glucose levels**.

### Control of blood glucose levels

Glucose is the main source of energy for respiring cells. It is the only source used by brain cells. After a meal:

> blood glucose levels rise

> cells in the pancreas detect the rise and secrete insulin into the blood

> insulin acts on liver cells

> liver cells convert glucose into **glycogen**

> glycogen molecules are stored in granules in liver cells

> blood glucose levels fall

> cells in the pancreas detect the fall and secrete **glucagon**

> glucagon makes liver cells convert glycogen back into glucose.

So, blood glucose levels rise and fall within safe limits.

The conditions inside your body, including blood glucose levels, are its **internal environment**. This is regulated to keep cells working properly. Maintaining a constant internal environment is **homeostasis**.

If blood glucose levels stay too high, this is a symptom of **diabetes**. Other symptoms include increased thirst, frequent urination and weight loss. Too much glucose in the blood increases blood pressure. Long term, this leads to damage to the kidneys, retina and blood vessels.

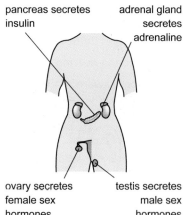

pancreas secretes insulin
adrenal gland secretes adrenaline
ovary secretes female sex hormones (such as oestrogen)
testis secretes male sex hormones (especially testosterone)

**FIGURE 2**: Four glands that secret hormones. Which one secretes insulin?

### QUESTIONS

**1** What do hormones do, in your body?

**2** Why will blood glucose rise after a meal but fall between meals?

**3** Explain the difference between glucose and glycogen.

**4** Why are high blood glucose levels harmful?



Q ... gcse hormones and glands

# Controlling diabetes

In **Type 1 diabetes** the pancreas does not make enough insulin. In **Type 2 diabetes** liver cells do not respond to insulin. In both cases, blood glucose levels rise too high.

High blood glucose levels can be detected in a drop of blood, using a blood glucose meter. Glucose appears in the urine only if blood levels are high. A simple dipstick test is used to test urine for glucose.

Diabetes cannot be cured, but it can be controlled. All Type 1 patients can inject insulin to lower their blood glucose levels. They monitor levels using a blood glucose meter and inject insulin, if levels are rising, or eat, if they are falling.

Type 2 diabetes is more likely in the obese. It is becoming more common, even in children. Diet, to control body mass and exercise, are sometimes enough to control Type 2 diabetes.

**FIGURE 3**: Insulin injections lower blood glucose levels. Insulin cannot be taken as a tablet because it is a protein and would be digested.

## Did you know?

It is a myth that eating too many sweets or sugary foods causes diabetes. However, obesity increases the risk of diabetes.

## QUESTIONS

**5** Explain the difference in treatment of Type 1 diabetes and Type 2 diabetes.

**6** How can a diabetic check their blood glucose levels?

# Negative feedback

Conditions inside the body are kept within narrow limits: any change automatically causes the body to adjust. It reverses that change. This effect, opposing a change, is called **negative feedback**. It brings the body back to normal.

Keeping blood glucose levels within safe limits is just one example of homeostasis controlled by negative feedback.

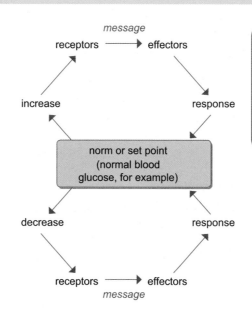

## QUESTIONS

**7** Redraw the diagram to summarise the structures and the changes in the control of blood glucose levels.

# Temperature control

**THE SCIENCE IN CONTEXT** To work effectively, your body temperature needs to stay in the range 36 to 37.5 °C. To stay healthy, the body must keep itself within this temperature. Measuring a person's temperature is one of the most frequently used health checks. Temperatures outside the range 36 to 37.5 °C suggest that something is wrong.

## Essential notes

> body temperature is controlled by the thermoregulatory centre in the brain

> heat loss is increased through increased sweating and increased blood flow in capillaries near the skin surface

> heat loss is reduced by decreased sweating and decreased blood flow in the capillaries near the skin surface

### Hypothermia

Normally, the temperature of internal organs is kept at about 37 °C. Hypothermia is a life-threatening condition, when the body's core temperature falls below 35 °C.

**FIGURE 1**: Warm clothing is essential in very cold conditions.

## Body temperature

### Why control body temperature?

**Enzymes** control the rates of chemical reactions in cells. A high temperature damages enzymes. A low temperature slows down reactions.

It is very important that body temperature is kept within narrow limits. Temperature control is called **thermoregulation**.

The body is kept at the temperature at which enzymes work best. A change in body temperature of 1 °C causes a 10% change in the rate of the enzyme-controlled chemical reactions.

### Heat loss

Temperature control mainly involves regulating the rate at which heat is being lost from the body. If temperature rises, more heat is lost. If temperature falls, less heat is lost.

In the brain, near where the spinal cord joins, is a **thermoregulatory centre**. This controls body temperature. It contains **thermoreceptors**, which detect changes in the temperature of its blood supply.

If blood temperature rises above 37 °C, heat loss from the body is increased by:

> dilating (widening) blood vessels in the skin – blood flow, through capillaries near the surface, is increased

> increasing sweating – **evaporation** of water, from the skin surface.

If blood temperature falls below about 37 °C, heat loss from the body is reduced by:

> constricting blood vessels in the skin – blood flow, through capillaries near the surface, is reduced

> decreasing the amount of sweating.

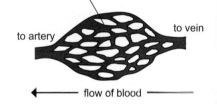

**FIGURE 2**: Capillaries carry blood close to the skin from veins to arteries.

### Did you know?

The skin has a very large surface area – about two square metres – in contact with the outside world.

### ○ QUESTIONS

**1** Why is thermoregulation necessary?

**2** Name the part of the brain that controls body temperature.

**3** How can heat loss, from the body, be reduced?

# How heat is lost from the body

At rest, most of the heat generated in the body comes from the liver and heart. Heat may also be gained from the environment. Heat is lost in four ways.

| Heat transfer process | Heat loss from body at rest (%) |
|---|---|
| sweating | 22 |
| conduction | 3 |
| convection | 15 |
| radiation | 60 |

> Evaporation of **sweat** converts liquid water to water vapour (gas). This requires energy. The energy is transferred from your body to the water molecules.

> Conduction is the **transfer** of energy from the body to a cooler object, when they are in contact.

> Convection is transfer of energy through air currents. Hot air rises and is replaced by cooler, more dense air. This is warmed and rises, causing **convection currents**.

> Radiation occurs when an object emits **infrared radiation**. All objects **emit** infrared radiation. The bigger the difference between the temperature of your skin and its surroundings, the faster energy is transferred.

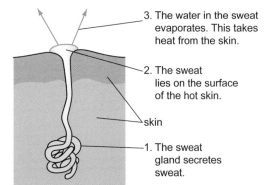

3. The water in the sweat evaporates. This takes heat from the skin.

2. The sweat lies on the surface of the hot skin.

skin

1. The sweat gland secretes sweat.

**FIGURE 3**: How sweating cools you down.

**FIGURE 4**: Why does this athelete need to lose more heat than spectators?

## QUESTIONS

**4** Name four ways in which your body can lose energy.

**5** Why is sweating an effective method for cooling the body?

**6** Explain the difference between dilation and constriction of skin capillaries.

# Negative feedback

Just like blood glucose levels, body temperature is regulated by a negative feedback mechanism.

A rise in body temperature triggers mechanisms which automatically bring the temperature down again. A fall in temperature triggers the mechanisms which cause it to rise again. So body temperature falls and rises slightly, between about 36.5 °C and 37.5 °C, but always returns to 37 °C.

## QUESTIONS

**7** Suggest why an excessive increase in body temperature might go out of control through positive feedback.

# Preparing for assessment: Planning a practical investigation

To achieve a good grade in science, you not only have to know and understand scientific ideas, but you need to be able to apply them to other situations and investigations.

Connections: This task relates to Unit 2 context 3.4.1.1 Control of body systems. Working through this task will help you to prepare for the practical investigation that you need to carry out for your Controlled Assessment.

## ✳ Investigating the body's temperature control

Harry carried out a practical to investigate some factors that affect the rate of heat loss through a person's skin. In the practical he used four boiling tubes.

> Tube A was uncovered to represent bare skin with no hairs on it.

> Tube B was loosely wrapped with a sheet of kitchen towel, held in place with sticky tape, to represent skin with hairs standing on end.

> Tube C was tightly wrapped with a sheet of kitchen towel, held in place with an elastic band, to represent skin where erector muscles are relaxed and the hairs are laying flat.

> Tube D was tightly wrapped with a sheet of kitchen towel, held in place with an elastic band and soaked in water, to represent the skin of a person sweating.

Harry's method was:

> standing the four boiling tubes in a rack and putting a thermometer into each

> measuring 40 cm³ of hot water into each boiling tube

> for the fourth boiling tube (Tube D) only, pouring 10 cm³ of hot water over the kitchen towel on the outside of the tube

> recording the temperature of the water in each tube every three minutes, for half an hour

> plotting graphs (cooling curves) of temperature against time.

His conclusion was that sweating cools the body and that increased insulation reduces heat loss.

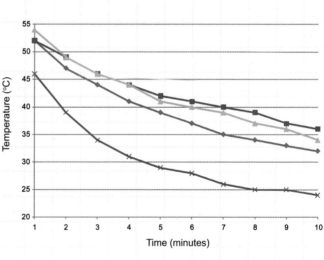

FIGURE 1: Harry's graphs.

## Investigating fabrics, heat loss and sweating

Harry is a keen hiker and cyclist and is interested in how different types of outdoor clothing can reduce heat loss. He saw an advertisement in one of his outdoor activity magazines for a T-shirt that keeps you warm, but does not trap sweat – it can 'breathe'.

He decided to plan an investigation of how different types of fabric affect heat loss and sweating.

## ✸ Research

**1.** Find out what fabrics are used to make outdoor clothing and T-shirts for sporting and outdoor activities. What claims do manufacturers make about the clothing? For example: waterproof, warm, breathable.

You need to find and use at least three research sources. These might be magazines, books, manufacturer's labels or the internet. Make sure that you list the sources of information you use.

**2.** Research the science behind what affects the properties of fabrics used in outdoor clothing and T-shirts for sports and outdoor activities.

Use this research to help you plan the investigation.

## ✸ Planning

**3.** You decide to work with Harry on this investigation. Say what you think the purpose of this investigation might be.

Make sure that the purpose is clear. Do not be too general – state some specific outcomes that he is looking for.

**4.** Write a summary of the procedure you would use (based on the method Harry used in his first practical and your research).

Give a broad outline and list equipment and materials that you will need. You need to identify the variables that affect the investigation.

**5.** Write your own step-by-step plan for the investigation.

The plan must be structured, with well-ordered steps, so that it can be followed by another person. Identify the independent variable (the one you will vary), the dependent variable (the one that the independent variable affects) and all the control variables (the other variables that will be kept constant). Remember to include the range of readings that you need to take, the time intervals between them and some guidance on how many readings.

**6.** Explain carefully a relationship between two variables that your plan would allow you to measure.

Describe how you will look for the trends and patterns in the relationship between the independent variable and the dependent variable.

## ✸ Assessing and managing risks

Make sure that you say what the risks are for each hazard you list.

**7.** Looking at your plan for the investigation, list any potential hazards and give the associated risks.

**8.** Explain how you would minimise the risks.

Include methods of control. Support these with the scientific reasoning for the steps you suggest to minimise risks.

# Body chemistry

**Essential notes**

> the body functions properly due to a series of complex chemical reactions

**THE SCIENCE IN CONTEXT** In your body, vast numbers of chemical reactions are happening all the time. They are what make your body function. Chemical reactions that happen during digestion and respiration are examples. Many scientists are interested in body chemistry, including healthcare workers, toxicologists, sports scientists and forensic scientists.

## Energetic sleep

You use energy to do all the things you do, from growing to thinking and from talking to walking. You even use energy when you are sleeping.

**FIGURE 1**: Food is the body's fuel. When you eat, it is like filling a car up with petrol. Lots of chemical reactions in your body release the energy from the food.

## Digestion and respiration

### Digestion

Chemical reactions do not happen only in test tubes. They are happening all the time in your body. **Digestion** is a good example.

The main food groups are:

> carbohydrates

> proteins

> fats.

These chemical compounds are broken down, in the digestive system, by **chemical reactions**. Chemicals, called enzymes, speed up these chemical reactions.

For example, chew a piece of bread for a while and you will find it tastes sweeter. This is because starch (a carbohydrate) in the bread is changed to sugars, by chemical reaction.

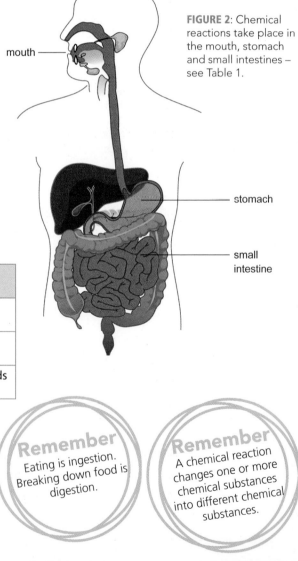

FIGURE 2: Chemical reactions take place in the mouth, stomach and small intestines – see Table 1.

mouth

stomach

small intestine

| Part of body | Chemical reaction(s) | | |
|---|---|---|---|
| mouth | carbohydrates → sugars | | |
| stomach | carbohydrates → sugars | proteins → amino acids | |
| small intestine | carbohydrates → sugars | proteins → amino acids | fats → fatty acids + glycerol |

**TABLE 1**: Chemical reactions happening in the body.

### Respiration

**Respiration** is not the same as breathing. Respiration is the chemical changes, in the body, that release energy from glucose.

Glucose is a **sugar** made during digestion. It is our main source of energy.

**Remember**
Eating is ingestion. Breaking down food is digestion.

**Remember**
A chemical reaction changes one or more chemical substances into different chemical substances.

A chemical reaction takes place between glucose and oxygen in cells. Different chemical substances are made – **carbon dioxide** and water – and energy is released. This is **aerobic respiration**. The word equation for the chemical reaction is:

glucose + oxygen → carbon dioxide + water + <u>energy</u>

Glucose and oxygen are the **reactants**. Carbon dioxide and water are the **products** of the reaction. Although energy is released, it is not a product. Only chemical substances are products.

## QUESTIONS

**1** Name the types of chemical compounds in the three main food groups.

**2** Where is starch chemically changed into sugars?

**3** Where does aerobic respiration happen?

## Enzymes

Chemical reactions in the mouth, stomach and small intestine need help from enzymes. Enzymes are catalysts. They make chemical reactions go faster. Enzymes are also quite choosy. Each one catalyses a specific type of chemical reaction.

| Chemical reaction | Enzymes that catalyse the reaction |
|---|---|
| carbohydrates → sugars | amylase |
| proteins → amino acids | pepsin |
| fats → fatty acids + glycerol | lipase |

The chemical changes are also helped by:

> hydrochloric acid produced in the stomach

> bile produced in the liver.

Enzymes in the stomach work better in acidic conditions. This is why the stomach produces hydrochloric acid.

Enzymes in the small intestine work better in alkaline conditions. Bile helps to provide these conditions.

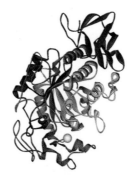

**FIGURE 3**: Amylase is an enzyme. This very large molecule is in saliva and helps you to digest carbohydrates.

## QUESTIONS

**4** Explain why making amino acids from protein is a chemical reaction.

**5** Give the chemical name for the acid in your stomach.

**6** What are the products of aerobic respiration?

## Aerobic and anaerobic

Oxygen is carried around the body in the blood. The blood carries it to cells that need energy, such as muscle cells. Given enough oxygen, the energy comes from aerobic respiration. However, during hard exercise, you cannot breathe in enough air to replace oxygen as quickly as it is being used up.

When this happens, the body turns to anaerobic respiration. It does not transfer as much energy, but it does not need oxygen. The chemical reaction is:

glucose → lactic acid + <u>energy</u>

Glucose is the reactant and lactic acid is the product.

Have you have ever felt intense pain in your legs when running hard? The reason is that lactic acid builds up in your muscles and your body cannot remove it quickly enough. Athletes that run 400 and 800 metres know all about it. It is extremely painful.

## QUESTIONS

**7** Describe how a runner obtains energy to run 400 metres in a fast time.

Q … aerobic and anaerobic respiration

# Acids and bases

## Essential notes
> acids and bases are hazards
> there are ways to minimise risks from using them

**THE SCIENCE IN CONTEXT** Scientists find it useful to group chemical substances according to their properties. Acids and bases are two examples. Safe working in the laboratory or in the field is vital. Whether they are acids, bases or other substances, risks from using hazardous materials must always be minimised.

**FIGURE 1**: The name formic acid comes from *formica*, Latin for ant, because it was first isolated by distilling ants.

## Bites and stings

Ever been bitten by an ant or stung by a bee? It can be painful and make the skin red and itchy. In both cases, the main culprit is the same – formic acid (modern name: methanoic acid).

## Acids, bases, hazards and risks

### Acids and bases

Acids and bases are two groups of chemical compounds.

A solution of an acid in water is **acidic**. Acids turn litmus red.

Bases that dissolve in water are called alkalis (the solutions are **alkaline**). Alkalis turn litmus blue.

An acid reacts with a base to produce a salt and water.

Acids and bases often found in the laboratory include:

> **Acids**
hydrochloric acid, HCl(aq)
sulfuric acid, $H_2SO_4$(aq)
nitric acid, $HNO_3$(aq)

> **Soluble bases (alkalis)**
sodium hydroxide, NaOH(aq)
potassium hydroxide, KOH(aq)
ammonia, $NH_3$(aq)

> **Insoluble bases (not alkalis)**
copper oxide, CuO(s)
zinc hydroxide, Zn(OH)2(s)

Note: (aq) indicates an aqueous solution (that is, a solution in water); (s) indicates a solid.

Acids and bases can be dangerous. What does 'dangerous' mean? Two words help to unpick the idea – **hazard** and **risk**.

### Hazards and risks

A hazard is something that *could* cause harm. Risk, on the other hand, is the *likelihood* of harm being done.

You probably do not think of lemon juice as dangerous, but it does have the potential to do harm – it hurts if it squirts into your eye. The risk of this happening depends on how careful you are. The risk is almost eliminated if you wear spectacles.

> **Remember**
> All alkalis are bases, but not all bases are alkalis.

### Did you know?
Your stomach contains hydrochloric acid. If it did not, your body would not be able to digest food properly.

## QUESTIONS

**1** Write a word equation for the reaction between an acid and a base.

**2** Describe the symbol that tells you a substance is toxic.

**3** Name the two hazards associated with vinegar (a dilute acid).

## Identifying hazards

You need to be able identify hazards and take steps to reduce the risk of harm being done. You will need to do this when planning an investigation.

To help identify hazards, each type has its own hazard warning symbol. Here are some important ones.

Corrosive    Flammable    Oxidising    Explosive    Harmful

Irritant    Toxic    Biohazard    Harmful to the environment    Radiation

## Minimising risk

Control measures can be put in place to minimise the risk of harm when using acids and alkalis. They are:

> wear eye protection (safety spectacles or goggles)

> wear protective gloves

> wear a laboratory coat or apron

> after use, always replace the stopper in the bottle containing the acid or alkali.

If for any reason the acid or alkali does get on the skin or in an eye, make sure there is access to:

> running cold water, so that the skin can be washed thoroughly

> eyewash, so that the eye can be washed thoroughly.

These last two actions do not reduce the risk. They are remedial actions to minimise any damage.

FIGURE 2: All laboratories have eyewash bottles. Some have full body showers. Many have fume cupboards.

### QUESTIONS

4 Why should you wear eye protection and protective gloves when working with acids and alkalis?

5 Having an eyewash bottle does not reduce the risk of using acids or alkalis. What is its purpose?

6 Suggest reasons why many chemical reactions are carried out in fume cupboards.

## Warning signs

Many household products contain acids or alkalis, as well as other ingredients. Acids and alkalis are irritants and corrosive. Other substances have different hazards associated with them.

If you search, you can find Material Safety Data Sheets (MSDS) for products. These list the ingredients, identify the hazards (sometimes only health hazards) and provide advice on, for example, first aid measures.

The active ingredient in many oven cleaners is sodium **hydroxide**, an alkali. It is very corrosive. Some similar products are available as aerosols that foam when sprayed on. The propellant is a substance such as butane. This is **flammable**.

### QUESTIONS

7 Investigate household products found in the kitchen, bathroom and garage. Describe the effects of each hazard if the product is used incorrectly.

Q ... acids bases hazards

# Neutralisation

**THE SCIENCE IN CONTEXT** Chemists are interested in chemical reactions: the way that one or more chemical substances change into different chemical substances. Just as with chemical substances, scientists find it useful to classify chemical reactions into a number of types. Neutralisation – the reaction of an acid with a base – is one of these.

## Taking away the pain

Bee stings are acidic. They have a pH of about 5.0–5.5. Calamine or sodium hydrogencarbonate (baking soda) can soothe the skin because both are slightly alkaline and may help to neutralise the acid.

### Essential notes

> in neutralisation reactions, an acid and base react to form a salt and water

> hydrogen ions ($H^+$) make solutions acidic and hydroxide ions ($OH^-$) make solutions alkaline

> neutralisation can be represented by: $H^+(aq) + OH^-(aq) \rightarrow H_2O(\ell)$

FIGURE 1: Calamine is a suspension of zinc oxide and some iron oxide in water.

## Neutralisation and pH

### Neutralisation

An acid reacts with a base to produce a salt and water. For example:

hydrochloric acid + sodium hydroxide
→ sodium chloride + water

$HCl(aq) + NaOH(aq) \rightarrow NaCl(aq) + H_2O(\ell)$

sulfuric acid + zinc oxide → zinc sulfate + water

$H_2SO_4(aq) + ZnO(s) \rightarrow ZnSO_4(aq) + H_2O(\ell)$

These are **neutralisation** reactions.

### pH

pH can be estimated using indicator solution or paper. An accurate value can be obtained using a pH meter.

> A neutral solution has a pH of 7.

> An acidic solution has a pH less than 7.

> An alkaline solution has a pH greater than 7.

#### Did you know?

Many people say that wasp stings are alkaline and should be treated with vinegar. In fact, they are almost neutral (pH 6.8–6.9).

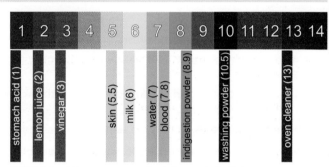

FIGURE 2: pH is a measure of acidity and alkalinity.

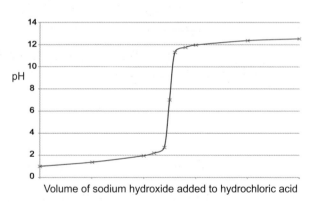

FIGURE 3: As sodium hydroxide solution is added to hydrochloric acid, the pH changes sharply. The acid has been neutralised.

## QUESTIONS

1 Name the salt produced when hydrochloric acid reacts with zinc oxide.

2 Name the salt produced when sulfuric acid reacts with copper oxide.

3 What would a scientist use to measure pH accurately?

# Hydrogen ions

Acidic solutions (pH < 7) are acidic because there are hydrogen ions in solution. For example, hydrogen chloride is a gas, $HCl(g)$. When hydrogen chloride gas dissolves in water, it produces hydrochloric acid, $HCl(aq)$.

$$HCl(g) + aq \rightarrow HCl(aq)$$

$HCl(aq)$ is chemical shorthand for $H^+(aq) + Cl^-(aq)$.

You could write:

$$HCl(g) + aq \rightarrow H^+(aq) + Cl^-(aq)$$

Alkaline solutions (pH > 7) are alkaline because there are hydroxide ions in solution. For example, sodium hydroxide is a solid, $NaOH(s)$. When solid sodium hydroxide dissolves in water, it produces sodium hydroxide solution, $NaOH(aq)$.

$$NaOH(s) + aq \rightarrow NaOH(aq)$$

$NaOH(aq)$ is chemical shorthand for $Na^+(aq) + OH^-(aq)$.

You could write:

$$NaOH(s) + aq \rightarrow Na^+(aq) + OH^-(aq)$$

## Neutralisation of ions

In a neutralisation reaction, hydrogen ions react with hydroxide ions to produce water molecules.

| $HCl(aq)$ | + | $NaOH(aq)$ | $\rightarrow$ | $NaCl(aq)$ | + | $H_2O(\ell)$ |
|---|---|---|---|---|---|---|

Writing the ions present in each case:

| $H^+(aq) + Cl^-(aq)$ | + | $Na^+(aq) + OH^-(aq)$ | $\rightarrow$ | $Na^+(aq) + Cl^-(aq)$ | + | $H_2O(\ell)$ |
|---|---|---|---|---|---|---|

The ions shown in blue do not change and are not involved in the reaction. Sometimes called 'spectator ions', they are just there watching what else is going on. Omitting them, the equation is simply:

| $H^+(aq)$ | + | $OH^-(aq)$ | $\rightarrow$ | $H_2O(\ell)$ |
|---|---|---|---|---|

This ionic equation applies to all neutralisation reactions between an acid and an alkali.

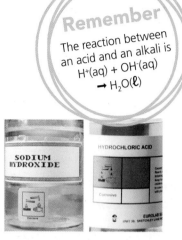

**Remember**

The reaction between an acid and an alkali is $H^+(aq) + OH^-(aq) \rightarrow H_2O(\ell)$

**FIGURE 4**: These two hazardous solutions react to produce a solution of sodium chloride – a compound that is a vital part of your diet.

## QUESTIONS

**4** Which ions are always present in (a) acidic solutions (b) alkaline solutions?

**5** Write equations to show the ions present in solution when sulfuric acid, $H_2SO_4(\ell)$ dissolves in water.

**6** Show why the reaction between sulfuric acid and potassium hydroxide solution can be represented as $H^+(aq) + OH^-(aq) \rightarrow H_2O(\ell)$.

# Dilute, concentrated, weak and strong

The *concentration* of an acid describes how much of it is dissolved in a known quantity of solution.

The *strength* of an acid, however, is a measure of the concentration of hydrogen ions in the solution.

A strong acid ionises completely in water. Hydrochloric acid is a strong acid. A weak acid only partially ionises. Ethanoic acid is a weak acid.

## QUESTIONS

**7** Explain why 0.1 mol/dm³ hydrochloric acid reacts with magnesium faster than 0.1 mol/dm³ ethanoic acid does.

| Acid | Particles present in solution | pH of 0.1 mol/dm³ solution |
|---|---|---|
| hydrochloric acid | $H^+(aq)$, $Cl^-(aq)$ | 1 |
| ethanoic acid | $CH_3COOH(aq)$, $CH_3COO^-(aq)$, $H^+(aq)$ | 3 |

# Stomach acid

**Essential notes**

> the stomach works most effectively in acid conditions

> an antacid neutralises excess stomach acid to help to treat heartburn and stomach ulcers

**THE SCIENCE IN CONTEXT** Hydrochloric acid in your stomach is needed to break down protein you eat and to help protect you from harmful microorganisms in your food. However, excess acid can sometimes cause heartburn and nausea. Antacids are used to control stomach acid. Pharmacologists investigate how drugs such as antacids work so that they can be used effectively and safely.

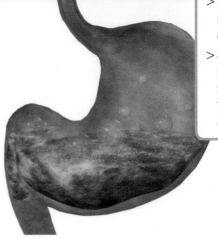

## Sticky mucus

Your stomach secretes a sticky mucus that clings to its walls. If the mucous layer is damaged, acid gets through and can result in painful ulcers.

**FIGURE 1**: Your stomach is protected by a thin membrane which stops your stomach being (a) digested by its own enzymes (b) attacked by corrosive hydrochloric acid.

## Stomach acid

### The stomach and digestion

When you eat, your teeth grind food and increase its **surface area**. This provides a greater area for salivary amylase to start digesting starch. The food passes down the oesophagus to the stomach.

The stomach is about pH 2, owing mainly to hydrochloric acid that is produced by cells in the stomach wall. pH 2 means that the concentration of hydrogen ions is 0.01 mol/dm³. The acid has three important functions.

> It helps continue the process of digestion.

> It kills many harmful microorganisms (not all – some that cause food poisoning have adapted to survive at low pH).

> It provides acidic conditions (needed to help protease enzymes break down proteins).

### Too much acid

Too much acid can cause problems such as indigestion and stomach ulcers. Antacids are substances used to neutralise excess acid. They are bases that react with the excess acid. An example is **calcium carbonate**:

calcium carbonate + hydrochloric acid →
calcium chloride + carbon dioxide + water

$$CaCO_3(s) + 2HCl(aq) \rightarrow CaCl_2(aq) + CO_2(g) + H_2O(\ell)$$

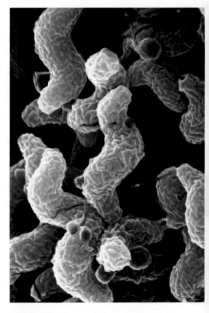

**FIGURE 2**: Campylobacter is a microorganism that has adapted to survive at low pH.

## QUESTIONS

1 Give the pH of stomach acid.

2 Where is hydrochloric acid produced in the stomach?

3 Give an example of an antacid.

### Did you know?

An empty stomach has a volume of about 40 cm³. Normally it expands to hold about 1000 cm³ of food, but can expand further, up to 4000 cm³.

# Antacids

Antacids are a group of drugs that relieve:

> indigestion (dyspepsia)

> heartburn (inflammation of the oesophagus caused by acid refluxing up from the stomach – acid reflux)

> some tpes of nausea.

Modern drugs are more effective than antacids. Antacids simply neutralise the acid for a short period, but drugs developed recently reduce the amount of acid made by the stomach.

Antacids are 'over the counter' (OTC) medicines. This means that you do not need a doctor's prescription to get them. They come as tablets, powders and liquids. The active ingredients are bases that neutralise excess stomach acid.

Here are some of the active ingredients used in antacids and the equations for their reactions with hydrochloric acid:

> calcium carbonate

$$CaCO_3(s) + 2HCl(aq) \rightarrow CaCl_2(aq) + CO_2(g) + H_2O(\ell)$$

> magnesium hydroxide

$$Mg(OH)_2(s) + 2HCl(aq) \rightarrow MgCl_2(aq) + 2H_2O(\ell)$$

> aluminium hydroxide

$$Al(OH)_3(s) + 3HCl(aq) \rightarrow AlCl_3(aq) + 3H_2O(\ell)$$

> sodium bicarbonate

$$NaHCO_3(aq) + HCl(aq) \rightarrow NaCl(aq) + CO_2(g) + H_2O(\ell)$$

**FIGURE 3**: What is an OTC medicine?

## QUESTIONS

**4** If calcium carbonate was added a little at a time to some dilute hydrochloric acid, how would you know when the acid had been neutralised?

**5** You might have noticed that none of the antacids are alkalis. Why do you think this is?

# Comparing antacids

Antacids come in various forms, with differing active ingredients and formulations, and differing price tags. How could this effectiveness be compared?

Human trials would provide the best evidence. These are costly, if carried out properly. For example, large sample and control groups would be needed.

Some useful initial information could come from working on a 'model' stomach. 0.01 mol/dm³ hydrochloric acid at 37 °C (body temperature) would be a suitable model for stomach acid.

A measured mass of the antacid would be added to a measured volume of 0.01 mol/dm³ hydrochloric acid at 37 °C. The change in pH could be monitored using a pH meter. A graph of pH against time would indicate the rate of reaction and the extent to which the antacid raised the pH. If carbon dioxide is produced in the neutralisation, the rate of reaction might be determined, by recording the loss in mass of the reaction mixture or the volume of carbon dioxide produced.

## QUESTIONS

**6** Based on the outline, left, write a procedure for comparing antacids.

# Developing new drugs

**Essential notes**
> new drugs must be fully tested before they are approved
> testing new drugs on animals and humans raises some issues

**THE SCIENCE IN CONTEXT** Developing new drugs is time-consuming and expensive. It involves extensive testing and trialling. The process involves teams of scientists with various specialist knowledge and skills. They include computer modellers, synthetic and analytical chemists, biochemists and pharmacologists. Technologists and engineers also contribute by developing the manufacturing process.

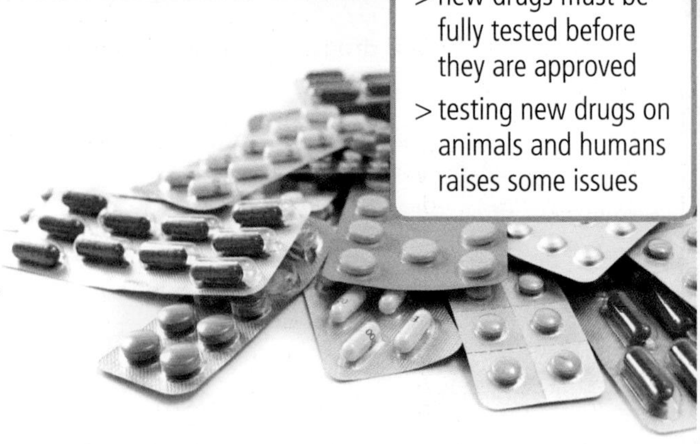

## UK pharmaceutical industry

It can take 10 to 15 years to develop a new medicine, at an average cost of £600 million. The disease to be treated is identified and 5000 to 10 000 compounds are investigated.

**FIGURE 1**: Each year, the UK pharmaceutical industry markets around 20 new medicines.

## Medicines and testing

### Body chemistry and medicines

**Illness** and **disease** happen when chemical reactions in your body do not happen as they should. Medicines often help.

Most medicines are **formulations**. A formulation is the recipe for making a medicine:

> quantities of ingredients
> how to combine them
> their form (such as tablets or ointment)
> how to administer them.

The active ingredients are medical **drugs** (also called therapeutic drugs).

### Extensive testing

Any new medical drug is tested extensively using:

> experiments with living cells
> chemical tests
> computer modelling
> animal testing
> clinical trials on human volunteers.

No medical drug has zero risk. Before it is approved for use, its benefits must be shown to outweigh its risks.

**Remember**
Drugs are substances that change chemical reactions in the body.

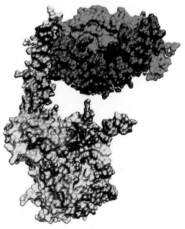

**FIGURE 2**: This computer image shows how an anti-cancer drug (green) can attach to a protein (red) and prevent cancer cells from multiplying.

## QUESTIONS

**1** What is the relationship between disease and chemical reactions in the human body?

**2** What is a formulation?

**3** Suggest why clinical trials on humans are the final tests to be carried out.

### Did you know?

Stomach acid destroys some medicines. Tablets are coated with a substance that stops them dissolving until they reach the intestine.

Q ... gcse drug testing

## Discover and develop

It can take 10 to 15 years to discover and develop a new drug.

Scientists start by trying to understand the disease and its causes. They choose a **target** for the potential new medical drug. This might be a molecule in the body, such as a gene or a protein.

▼

**Candidate drugs** that will act on this target (and alter the course of the disease) are made. Their safety and effectiveness is tested (**preclinical testing**), though not on human volunteers.

The most promising candidate drug is chosen for further study. This can take three to six years.

▼

An application is made to conduct **clinical trials** using human volunteers. These can take six to seven years.

Phase 1 trials involve 20–100 healthy volunteers. Phase 2 and 3 trials involve, respectively, 100–500 and 1000–5000 volunteer patients that have the disease.

▼

The final one to two years are spent reviewing data, to decide if the drug can be approved for use.

**FIGURE 3**: Clinical trials use human volunteers.

### Tests using non-human animals

Testing on non-human animals is a required part of any drug development programme. It provides vital information that cannot be obtained by other means.

> **Remember**
> The UK has strict regulations about animals for medical testing.

**For:** Animal testing has allowed the development of, for example, cancer and HIV drugs, insulin, antibiotics and vaccines. Most scientists believe that animal testing provides valuable data relevant to humans.

**Against:** Many animals used for testing live their lives in captivity and are killed after their use. Also, the way a drug behaves in a non-human animal is different from its behaviour in a human: some people believe animal testing is unreliable.

These arguments can be overshadowed by ethical considerations. Is it 'right' to use animals in this way? Emotions play a big part.

### QUESTIONS

**4** Explain the link between a 'target' and a 'candidate drug'.

**5** Explain the purpose of preclinical testing.

**6** Summarise your own views on animal testing.

## Designing clinical trials

Planning and carrying out trials properly is essential. Bias must be eliminated. To avoid bias, many trials use the following protocols.

> Some volunteers receive the candidate drug while others are given the standard treatment or a **placebo** (a treatment with no therapeutic effect).

> Each volunteer is randomly assigned to the candidate drug or the standard treatment (or placebo).

> Neither the researchers nor the volunteers know which treatment is being given until the study is complete (this is called a double blind trial).

The number of volunteers is an important factor. The more that are involved, the greater the statistical significance of the data. This makes the trial more expensive and difficult to undertake.

### QUESTIONS

**7** Explain how the trials, described on the left, help to eliminate bias and increase the reliability of data gathered.

# Chromosomes, genes and DNA

... gcse DNA

**THE SCIENCE IN CONTEXT** Cells are your body's chemical factories. They are where the chemistry happens. Cells combine to make tissues, tissues make organs and organs make up you. Inside each of your cells is a nucleus. It contains chromosomes and genes. These are inherited from your parents and determine your characteristics. Scientists who study genes are called geneticists.

## Watson and Crick

In 1953, James Watson and Francis Crick described the structure of the DNA molecule. They demonstrated this structure in a model made out of clamps and card. It was based on two spirals forming a double helix.

### Essential notes

> each chromosome in the nucleus of a cell contains large numbers of genes
> genes control the characteristics of the body

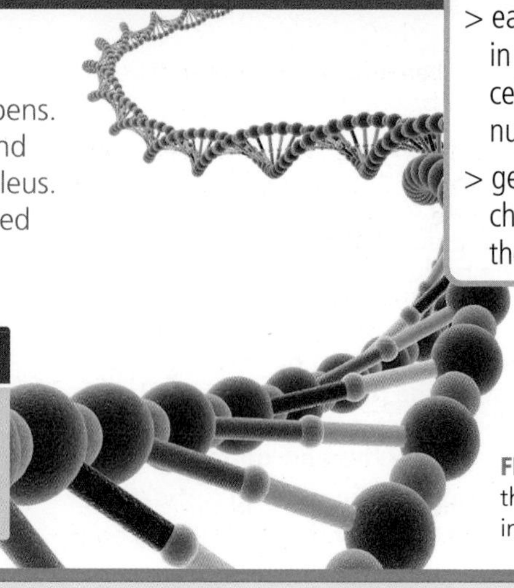

**FIGURE 1**: DNA is the molecule found in chromosomes.

## Your cells

Like all animals, you are made up of cells. They clump together to make tissues. Tissues combine to make organs such as your heart, liver and kidneys.

People tend to look like their parents. This is because information about a whole range of characteristics is passed on from parents to offspring in **genes**. These genes are carried in gametes (sex cells). Male gametes are **sperm** cells and female gametes are **egg cells**.

Of course, not all aspects of a person's appearance have been inherited from their parents.

### Chromosomes, genes and DNA

Your cells contain a **nucleus**. The nucleus is embedded in cytoplasm inside a cell membrane.

Inside the nucleus are **chromosomes**.

Chromosomes contain large numbers of genes. Chromosomes control how cells grow and what they do.

Genes are pieces of **DNA**. DNA is a large molecule with two helixes – spirals – twisted together. It is often described as a double helix.

DNA contains the genetic code.

The genetic code is the information that controls what happens in cells.

Our genes are inherited as DNA from our parents.

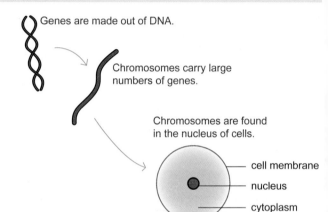

Genes are made out of DNA.

Chromosomes carry large numbers of genes.

Chromosomes are found in the nucleus of cells.

cell membrane
nucleus
cytoplasm

**FIGURE 2**: Each human chromosome carries a large number of genes made out of DNA.

### Remember

Every chromosome in the nucleus of a cell contains large numbers of genes made out of DNA.

## QUESTIONS

1 Where are genes found?

2 What are genes made out of?

3 Suggest why knowing about the structure of DNA is important – how could scientists use this information?

### Did you know?

Your body consists of about 10 000 000 000 000 cells. A line of 500 cells would be one centimetre long. You need a microscope to see them.

## Inheriting genes

New cells are made by cell division. When this happens, **genes** can be copied and passed on.

A gene is a piece of DNA. During cell division, the two halves of a DNA double helix separate and each makes a new DNA molecule. Both are identical to the original one.

A chromosome contains many genes. There are 23 kinds of human chromosomes. Body cells contain two sets of chromosomes:

> one set from the mother's egg cell

> the other set from the father's sperm cell.

These sex cells (gametes) join together during **fertilisation**, so that we get half of our genes from each parent.

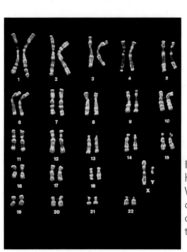

FIGURE 3: The full set of human chromosomes. Why are these chromosomes made out of two halves joined together?

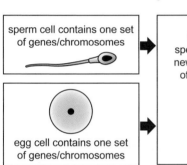

sperm cell contains one set of genes/chromosomes

egg cell contains one set of genes/chromosomes

fertilisation: egg and sperm fuse together – the new cell now has two sets of genes/chromosomes

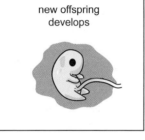

new offspring develops

FIGURE 4: From egg to offspring. How many chromosomes will you find in a fertilised human egg cell?

### QUESTIONS

**4** Why is it necessary for DNA to be able to copy itself?

**5** What are gametes?

**6** There are about 25 000 human genes. Roughly how many genes are there on each chromosome?

## How DNA works

Genes control cell activities by controlling the proteins that the cell makes. Some of the proteins are used to make structures inside cells. Other proteins are enzymes. Enzymes control the chemical reactions going on inside cells.

When a cell makes proteins, it joins up **amino acids** in a definite order. Cells are able to make many different proteins using the different DNA codes in the genes.

Sometimes the code goes wrong and a **protein** cannot be made correctly. Sometimes this is harmless; for example blood group O does not contain the factor A or B found in other blood groups. Sometimes the code can cause problems. For example, the protein in haemoglobin molecules does not form correctly in people who have sickle cell anaemia.

### Did you know?

DNA stands for deoxyribonucleic acid.

### QUESTIONS

**7** How is it possible for genes to control chemical reactions inside cells?

# Variation

**THE SCIENCE IN CONTEXT** Variety is the spice of life. It is also a characteristic of living organisms. No two human beings are identical, not even identical twins. Identical twins have virtually the same DNA, but they have different fingerprints. Differences (scientist call them variations) between people may be genetic or environmental.

## Potato blight

Potato plants are grown from potatoes. Over time, a single original plant can give many thousands of genetically identical new ones. Unfortunately, the 'lumper' potato, in Ireland, was very susceptible to potato blight fungus disease. Between 1845 and 1852 the crop was destroyed. Around one million people died and a further million people emigrated.

## Causes of variation

**Variation** is the differences that you can see between organisms – between members of different species or of the same species.

Different sperm and egg cells contain different combinations of the parents' genes. This causes the differences between brothers or between sisters.

People's characteristics may be due to **genetic effects** or **environmental effects** or the **combined effects** of both. Here are some examples.

Genetic characteristics:

> eye colour

> hereditary disorders such as cystic fibrosis.

Environmental characteristics:

> scars

> tattoos.

Combined genetic and environmental effects:

> Genes decide how tall you could grow, but your actual height depends on your diet.

> Development in the womb is harmed by drugs, such as alcohol or nicotine, or disease such as German measles.

> Abilities, such as passing a science exam or mending a motorcycle, depend on learning.

### Twins

> Identical twins form from the same fertilised egg cell. They have identical sets of genes.

> Non-identical twins develop from separately fertilised egg cells. They share some genes, but can have many differences.

So, differences between identical twins are due to environmental effects. Sometimes identical twins are separated at birth and environmental effects become obvious. Different diets may cause one twin to be, for example, much thinner or shorter than the other.

FIGURE 2: You can tell your friends apart even though they belong to the same species. They have unique characteristics which help you to identify them.

FIGURE 3: Twins are useful for studying the causes of variation.

Q ... gcse variation

## QUESTIONS

1 Why are identical twins usually very alike?

2 Explain what the differences between identical twins can reveal.

3 Which of these human characteristics are determined by genetic effects, environmental effects or a combination: hair colour, shoe size, weight, blood group?

**Did you know?**

The children of the first American settlers grew up to be taller than their parents, because they had much more protein in their diet.

## Types of variation

### Discontinuous variation

In discontinuous variation, individuals have a definite characteristic. There are no 'in betweens'. Usually:

> there is a single gene

> the gene provides a code so a particular molecule can be made, or

> the gene is faulty so a molecule cannot be made.

An example is eye colour:

> for brown eyes the gene codes to make melanin in the iris

> for blue eyes the code is incorrect so melanin is not made.

### Continuous variation

Characteristics such as height show a gradual change from one extreme to another. This is continuous variation. Usually:

> several genes are involved

> there are a variety of 'in betweens'

> environmental factors may play a part.

An example is human skin colour. There are thought to be six genes involved. A continuous change from fair to dark is possible because of the tanning action of the Sun.

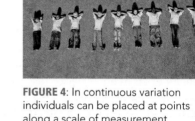

**FIGURE 4**: In continuous variation individuals can be placed at points along a scale of measurement.

## QUESTIONS

4 Say which of these characteristics show continuous variation and which show discontinuous variation (a) weight (b) eye colour (c) hair colour (d) length of forefinger (e) fused or free ear lobes?

## Variation within and between

### Variation within a species

Variation within a species helps its survival, if conditions change. The **fungus** disease that struck the nineteenth century Irish potato crop would not have been so devastating if some potato plants were resistant to it. Over time, species may gradually evolve into new species, better suited to new conditions.

### Variation between species

Each new species is adapted to suit its particular way of life. By being different, living organisms can exploit the variety in environments. For example, birds with different beaks can use different food sources.

## QUESTIONS

5 If each species is adapted to a particular way of life, why do individuals show variation?

Q ... gcse variation types

# Genes

**THE SCIENCE IN CONTEXT** Inherited family characteristics depend on genes. Members of a family tend to share many genes and so have similar characteristics. Genes provide the codes that cause, for example, eyes to be blue or brown. Sometimes, characteristics skip a generation or are hidden for several generations. This is because the effect of some genes may be masked by others.

## The Human Genome Project

The Human Genome Project is a map of the DNA molecules in 25 000 genes in human chromosomes. Scientists from around the world have been working on the project since 1990.

FIGURE 1: Genes provide the codes that cause eye colour.

## Genes and alleles

### Genes and alleles

Genes are pieces of DNA in chromosomes. Alternative forms of a gene are called **alleles** (pronounced al-eels). For example, the gene for ear lobe shape exists as an allele for fused lobes or an allele for free lobes.

Gametes are sex cells. Unlike body cells, which carry two sets of genetic information, gametes only contain one set of genetic information. When a male sperm cell fertilises an egg cell, one set of the mother's chromosomes combines with a set of the father's.

There is one allele for every gene from both the mother and the father. So, in offspring, the alleles occur in pairs.

### Mendel's pea experiments

About 150 years ago, Gregor Mendel carried out breeding experiments with tall and short pea plants. He fertilised plants by taking pollen grains from one and dusting them onto the stigma of another – a method known as crossing. When he crossed:

> tall plants with tall plants, he obtained tall plants

> dwarf plants with dwarf plants, he obtained dwarf plants.

This is pure breeding. When he crossed:

> tall plants with dwarf plants, he obtained tall plants

> the new tall plants with each other, he got a 3:1 mixture of tall and dwarf plants.

Mendel realised that whatever caused the dwarf characteristic must have remained intact. This is **monohybrid inheritance**. It is controlled by a single gene with two alleles.

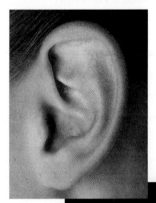

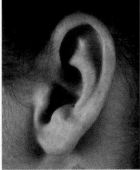

FIGURE 2: Fused and free ear lobes.

Q ... gcse genes ... gcse alleles

## Dominant or recessive alleles

Mendel thought tallness was caused by a dominant factor and dwarfness by a recessive factor. Mendel's 'factors' are called alleles today. A dominant allele paired with a recessive allele gives the dominant characteristic.

A dominant allele is represented by a capital letter; for example, T for tall. The recessive dwarf allele is lower case; t for example.

> Pure breeding tall plants are TT.

> Pure breeding dwarf plants are tt.

> In pure breeding tall plants, gametes T + T give TT: tall plants only.

> In pure breeding dwarf plants, gametes t + t give tt: dwarf plants only

> Tall TT crossed with dwarf tt combines gametes T + t to give Tt: tall plants, as T is dominant to t.

### QUESTIONS

**1** What happens to chromosomes that make alleles separate and pair?

**Remember**
An allele has a specific DNA code that produces a particular characteristic.

## Genetic diagrams

Genetic diagrams summarise how alleles behave.

| genotype | TT | x | tt |
|---|---|---|---|
| | pure breeding | | pure breeding |
| appearance | tall plant | | dwarf plant |
| gametes | T | | t |
| genotype | | Tt | |
| appearance | | tall | |

Offspring of pure breeding varieties are called **hybrids**. When Mendel crossed these new pea plants he obtained dwarf plants again. Punnett squares can be used to show how alleles can combine. They have columns for gametes from one parent and rows for the gametes from the other parent.

| genotype | Tt | x | Tt |
|---|---|---|---|
| appearance | tall | | tall |
| gametes | T or t | | T or t |

This Punnett square shows how dwarf plants reappear even though both parents are tall. It shows why Mendel obtained about three tall plants for every dwarf plant.

### QUESTIONS

**2** Why did Mendel find only tall plants when he crossed his original tall and dwarf plants?

## Punnett squares and likelihoods

Punnett squares show what happens 'on average' for many gametes.

In the example of the hybrid pea plants, the likelihood of obtaining a dwarf offspring is 1 in 4, or ¼ of all offspring. The likelihood of obtaining a tall offspring is 3 in 4, or ¾ of all offspring.

### QUESTIONS

**3** Catherine Zeta Jones does not have a cleft chin. Michael Douglas has a cleft chin caused by a rare dominant allele. Assuming that only one of Michael's parents has a cleft chin, what is the likelihood of the two actors having a child with a cleft chin?

# Inherited disorders

**THE SCIENCE IN CONTEXT** People inherit genes from their parents. Occasionally, faulty genes are inherited. They cause genetic disorders such as colour blindness and more serious conditions such as cystic fibrosis. Some disorders are more common than others. Geneticists are using their understanding of how genes work to try to improve treatments and develop cures for these genetic disorders.

## Gene therapy

Scientists have mapped the human genome and understand how genes work. They are trying to develop new techniques using 'normal' alleles to replace 'harmful' alleles.

### Essential notes

- > cystic fibrosis, sickle-cell anaemia, polydactyly and haemophilia are genetically inherited disorders caused by faulty alleles
- > Punnett squares can be used to work out the likelihood of a genetic disorder occurring
- > genetic screening can be used to test for the presence of harmful alleles in parents with a family history of a disorder, or their offspring

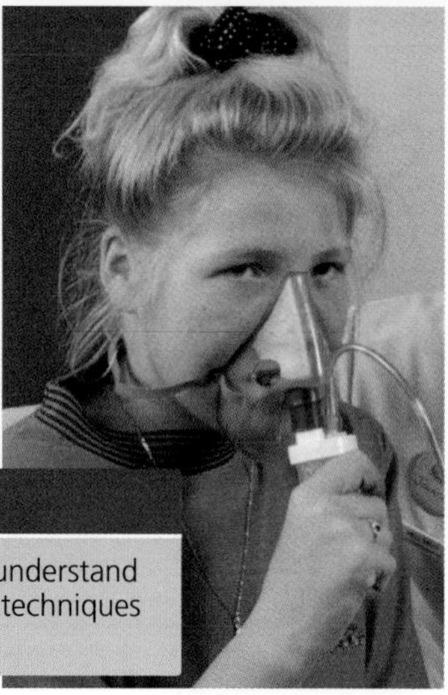

**FIGURE 1**: Cystic fibrosis causes breathing difficulties.

## Genetically inherited disorders

We inherit characteristics from our parents. Sometimes, the inherited characteristic is a disease. This is a **genetic disorder**.

The DNA code for a gene provides the instructions for making a protein. A change (**mutation**) in the DNA alters the code. This produces a faulty allele – it may be a recessive allele or a dominant allele.

**TABLE 1**: Some genetically inherited diseases.

| Inherited disorder | Allele |
|---|---|
| cystic fibrosis | recessive |
| haemophilia | recessive |
| sickle-cell anaemia | dominant |
| polydactyly | dominant |

To be born with a recessive disorder, you must get one recessive allele from each of your parents. They do not need to have the condition themselves. Parents who do not have a disorder, but can pass on the recessive allele that causes it, are called **carriers**.

Having a faulty allele means that cells:

- > are unable to make the protein, or
- > they make a faulty protein.

**Cystic fibrosis** causes thick and sticky mucus in lungs and airways, making breathing difficult. It can also affect sweat and digestive juices. Cystic fibrosis is caused by faulty recessive alleles. It is a recessive disorder.

**Haemophilia** is also caused by a faulty recessive allele. Haemophiliacs lack a protein needed to clot the blood. They risk bleeding to death. Fortunately haemophiliacs can be given the protein so that their blood clots normally.

**Remember**
Genetically inherited disorders are caused by faulty alleles.

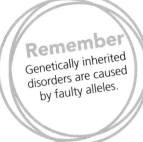

| | | Father | |
|---|---|---|---|
| | | C | c |
| Mother | C | CC | Cc |
| | c | Cc | cc |

**Key**

C is the cystic fibrosis allele

c is the normal allele

**FIGURE 2**: Two parents, each with a cystic fibrosis allele, have a one in four chance of having a child with cystic fibrosis.

### QUESTIONS

**1** Using F and f to represent the alleles, show how carrier parents without haemophilia can still pass the disorder on to their children.

**2** What is the likelihood that a child of carrier parents will be born with haemophilia?

## Polydactyly and genetic screening

### Polydactyly

**Polydactyly** means 'having many fingers'. It is a condition in which a person has more than five fingers on a hand, or more than five toes on a foot.

Polydactyly is caused by a dominant allele. This means that you only need to inherit one allele in order to have this condition. This family tree shows how polydactyly was inherited in one family.

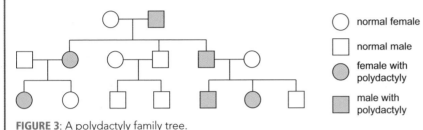

○ normal female

□ normal male

● female with polydactyly

■ male with polydactyly

**FIGURE 3**: A polydactyly family tree.

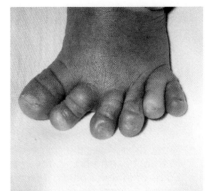

**FIGURE 4**: Polydactyly does not usually cause a person any problems.

### Genetic screening

Techniques used in the Human Genome Project (to identify human genes) can be used to check if normal or faulty alleles are present in an individual's DNA. If there is a family history of a condition, people can find out if they carry a faulty allele. It may help them to decide if they want to risk having a child who will suffer from the disorder.

### ● QUESTIONS

**3** A person had one grandmother with polydactyly. What was the likelihood that he also inherited the disorder?

**4** Suggest one advantage and one disadvantage of being screened for the haemophilia allele.

**5** What might cause someone to ask for genetic screening for faulty alleles?

### Did you know?

A harmless form of the cold virus is being used in trials to treat cystic fibrosis. The virus puts new human DNA into lung cells.

## Sickle-cell anaemia

Haemoglobin is the molecule that gives red blood cells their colour and helps them to transport oxygen.

**Sickle-cell anaemia**, a single gene recessive condition, is caused by a fault in the DNA code for part of the haemoglobin molecule. This causes red blood cells to become stiff and sickle shaped (crescent shaped) in low oxygen concentrations. The sickle-shaped cells block capillaries and reduce the blood supply to organs. This causes severe pain and organ damage. Victims may suffer strokes, heart attacks and kidney failure.

Carriers show no symptoms, but have a degree of **immunity** to malaria.

**FIGURE 5**: In sickle-cell anaemia, sickle-shaped blood cells can block capillaries.

### ● QUESTIONS

**6** Suggest why sickle-cell anaemia is more common in the descendants of people from areas where malaria was widespread.

Q ... huntington's disease ... sickle cell anaemia

# Preparing for assessment: Applying your knowledge

*To achieve a good grade in science, you not only have to know and understand scientific ideas, but you need to be able to apply them to other situations and investigations. These tasks will support you in developing these skills.*

## Neutralise or inhibit?

Antacids are over the counter (OTC) medicines used to treat heartburn and, for example, occasional bouts of indigestion. OTC means that you do not need a doctor's prescription to obtain them. Antacids are effective in providing short-term relief. However, they are not recommended for long-term use. For this, prescription medicines such as proton-pump inhibitors are more effective.

Antacids are usually available in tablet or liquid form and come in various different brand names. The active ingredients include calcium carbonate, aluminium hydroxide, magnesium carbonate and magnesium trisilicate.

Leaflets about medicines are often provided by doctors and at pharmacies. They provide information such as how the medicine works, who can and who should not take it, and possible side effects it may have. Most people who take antacids do not have any side effects. However, some may suffer from diarrhoea, constipation and belching.

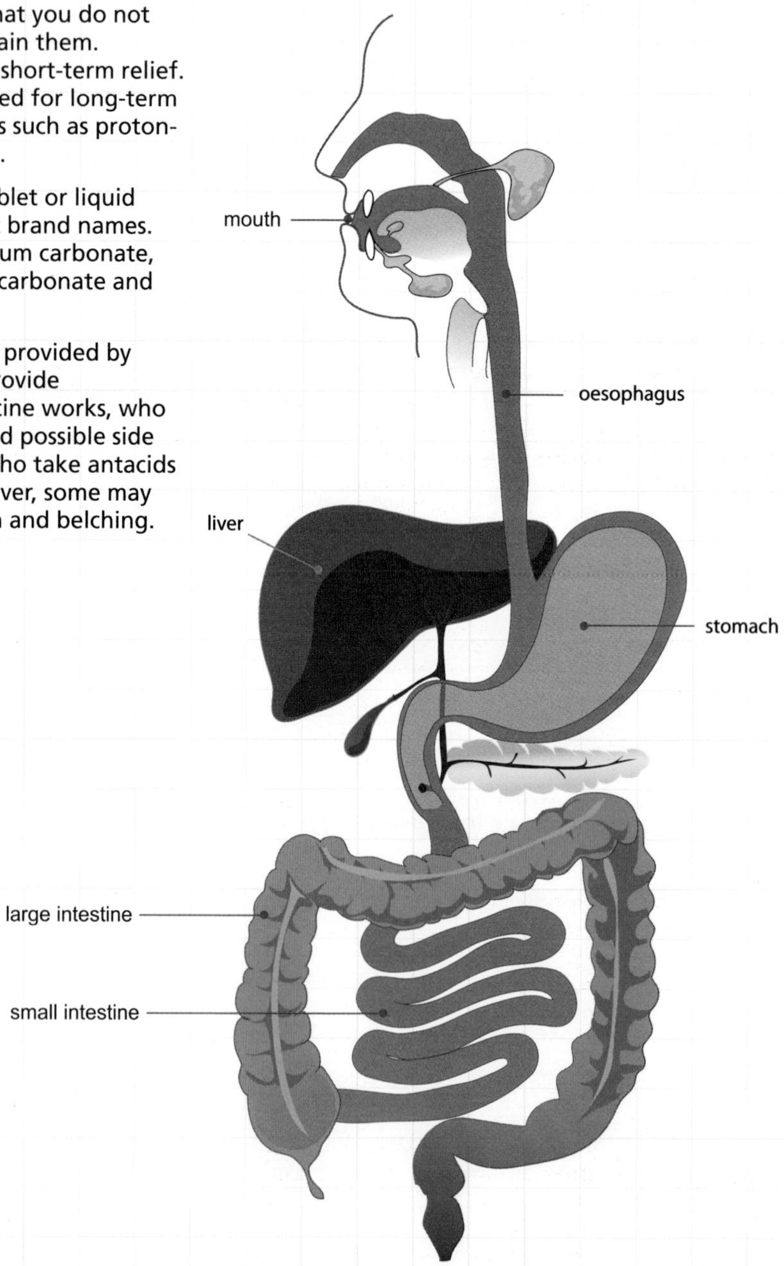

mouth
oesophagus
liver
stomach
large intestine
small intestine

## ☀ TASK 1

(a) Why are the contents of your stomach acidic?

(b) Although the term 'heartburn' is commonly used, in fact it has nothing to do with the heart. It is inflammation (reddening and soreness) of the oesophagus. What causes this inflammation?

(c) Indigestion occurs because there is too much acid in the stomach. What is the chemical name for this acid and what is its chemical symbol?

(d) What is the name of the process that occurs when you take an antacid such as magnesium carbonate or aluminium hydroxide? Try to find the chemical formula for these active ingredients.

(e) Write a word equation and a symbol equation for the reaction between calcium carbonate and stomach acid.

## ☀ TASK 2

(a) Suggest reasons why antacids come in tablet and in liquid form.

(b) Suggest a reason why some antacids are described as being liquids rather than solutions. Hint: Think about the solubility of compounds such as calcium carbonate, magnesium carbonate, magnesium hydroxide and aluminium hydroxide. Also, milk of magnesia is a liquid antacid. It gets the name 'milk' because it looks like milk – you cannot see through it – and you must shake it in the bottle before using it.

## ☀ TASK 3

(a) Acids contain hydrogen ions in solution. Write a balanced chemical equation for hydrogen chloride dissolving in water to form hydrochloric acid.

(b) Think about the structure of a hydrogen atom. What sub-atomic particles does a hydrogen ion, $H^+$, consist of?

(c) Modern prescription medicines for the relief of heartburn, indigestion and other digestive problems include proton-pump inhibitors. Explain how you think proton-pump inhibitors differ from antacids such as magnesium carbonate.

## ☀ TASK 4

Antacids containing magnesium compounds tend to be laxatives. Antacids containing aluminium compounds tend to be constipating. Give examples of these compounds and suggest why a number of commercial antacid preparations contain mixtures of magnesium and aluminium compounds.

## ☀ MAXIMISE YOUR GRADE

| Answer includes showing that you can... |
| --- |
| give one reason why the contents of people's stomachs are acidic. |
| name some of the compounds used as antacids and recognise that they are bases. |
| describe the reaction between acids and bases (oxides, hydroxides and carbonates) as neutralisation. |
| describe what makes a solution acidic and what makes a solution alkaline. |
| suggest reasons why medicines come in different forms. |
| write word equations and balanced symbol equations for chemical reactions. |
| explain how an antacid neutralises excess stomach acid to help to treat heartburn and nausea. |
| explain why medicines usually contain several ingredients. |

E

C

A

# Unit 2 Theme 1 Checklist

## To achieve your forecast grade in the exam you will need to revise

Use this checklist to see what you can do now. Refer back to the relevant pages in this book if you are not sure. Look across the three columns to see how you can progress. **Bold** text means Higher tier only.

Remember that you will need to be able to use these ideas in various ways, such as:

> interpreting pictures, diagrams and graphs
> applying ideas to new situations
> explaining ethical implications
> suggesting some benefits and risks to society
> drawing conclusions from evidence that you are given.

Look at pages 250–271 for more information about exams and how you will be assessed.

| To aim for a grade E | To aim for a grade C | To aim for a grade A |
|---|---|---|
| Recall examples of receptor cells that detect light, sound, smell, taste, touch and heat. | Describe how information from receptors passes along neurones in nerves to the brain. Know that the brain coordinates responses. | |
| Recall that reflex actions are automatic and rapid. | Know that reflex actions involve sensory, relay and motor neurones. | |
| Explain how longitudinal waves travel from vibrating objects to our ears for us to hear sounds. | Know that the human hearing range is 20–20 000 Hz. Know that homeostasis is how the body maintains a constant internal environment. | **Explain the principle of negative feedback in maintaining a constant internal environment.** |
| Know that hormones control many processes within the body and are secreted by glands into the blood to be transported to target organs. | Explain how the hormone insulin controls blood glucose levels. Know that high blood glucose levels are a symptom of diabetes. | |
| Recall that Type 2 diabetes may be controlled by changes in diet and exercise. Recall that Type 1 diabetes is controlled by insulin dosage. | Describe how the pancreas releases insulin into the blood to cause the liver to remove excess glucose from the blood and store it as insoluble glycogen. | **Describe how, if blood glucose concentration is too low, the pancreas releases glucagon, causing the liver to convert glycogen to glucose and release it into the blood.** |

## To aim for a grade E    To aim for a grade C    To aim for a grade A

| To aim for a grade E | To aim for a grade C | To aim for a grade A |
|---|---|---|
| Recall that the body maintains a constant temperature. | Describe how the thermoregulatory centre in the brain increases or decreases sweating, which cools the body by evaporation.<br><br>Describe how the thermoregulatory centre in the brain increases or decreases the blood flow to, and therefore the amount of heat lost from, the skin. | **Explain how negative feedback between an effector and the receptor of a control system reverses any changes to the system's steady state.** |
| Recall that the body functions correctly because of complex chemical reactions. | Name some hazards of acids and bases.<br><br>Describe some control measures that can be put in place to minimise risks from acids and bases. | Know that acids are neutralised by reaction with oxides, hydroxides or carbonates to form salts and other products. |
| Describe how a neutralisation reaction involves an acid and an alkaline substance reacting to form a salt and water. | Know that hydrogen ions ($H^+$) make solutions acidic.<br><br>Know that hydroxide ions ($OH^-$) make solutions alkaline. | **Know that neutralisation can be represented by the equation:**<br>$H^+ (aq) + OH^- (aq) \rightarrow H_2O(\ell)$. |
| Understand that the stomach works most effectively in acid conditions by helping to break down food. | Explain how an antacid neutralises excess stomach acid to help to treat heartburn and nausea. | |
| Recall that simple animal cells have a nucleus, cytoplasm and cell membrane.<br><br>Recall that the nucleus of a cell contains chromosomes. | Know that differences in the characteristics of individuals (variation) may be due to genetic causes or environmental causes or a combination of both. | |
| Recall that genes control the characteristics of the body. | Know that genes have different forms called alleles, which produce different characteristics.<br><br>Know that cystic fibrosis, sickle-cell anaemia, haemophilia and polydactyly are genetically inherited disorders. | Describe the mechanism of monohybrid inheritance using dominant and recessive alleles. |

## What you should know

### Rocks

Limestone is a sedimentary rock.

In the rock cycle, rocks are continually being broken down and then built up again.

 How do sedimentary rocks form?

### Energy

Energy can come from a variety of sources.

Energy can be transferred but cannot be used up.

Fossil fuels are finite, non-renewable sources of energy.

Burning fossil fuels produces pollutants such as carbon dioxide, which is linked to global warming.

 What is global warming?

### Sources of energy

Most of the electricity used at home is generated in power stations.

Certain alternatives to fossil fuels are renewable sources of energy.

List renewable sources of energy.

# You will find out

## Materials used to construct our homes

> Limestone is obtained from the ground and can be converted to quicklime and then into slaked lime.

> Limestone is used in the manufacture of cement and glass.

> The properties of metals, polymers and ceramics make them useful for construction.

> Polymers are manufactured using chemicals obtained from crude oil.

> Composites, such as reinforced concrete, MDF and GRP, are made from two or more materials.

## Fuels for cooking, heating and transport

> Fuels that can be used for cooking, heating and transport include natural gas, bottled gas, petrol, diesel, kerosene and heating oil.

> Burning fossil fuels produces pollutants such as sulfur dioxide.

> The heat produced from a hydrocarbon fuel is proportional to the number of carbon atoms in the molecule.

## Generation and distribution of electricity

> Electricity may be generated using fossil fuels, nuclear fuels and renewable energy resources to generate steam to drive turbines.

> The National Grid distributes electricity through cables from power stations to homes, businesses and industry.

# Limestone

**THE SCIENCE IN CONTEXT** Limestone is a rock that occurs naturally in many places. It is used in its natural state to construct buildings. It is a raw material for making cement, mortar, concrete and glass. Obtaining limestone raises the same issues as any operation that extracts materials from the ground.

### Common concrete

Concrete is the most used manufactured material in the world. The Egyptians used it building the pyramids. The Romans used it to construct buildings, bridges, aquaducts and other structures.

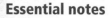

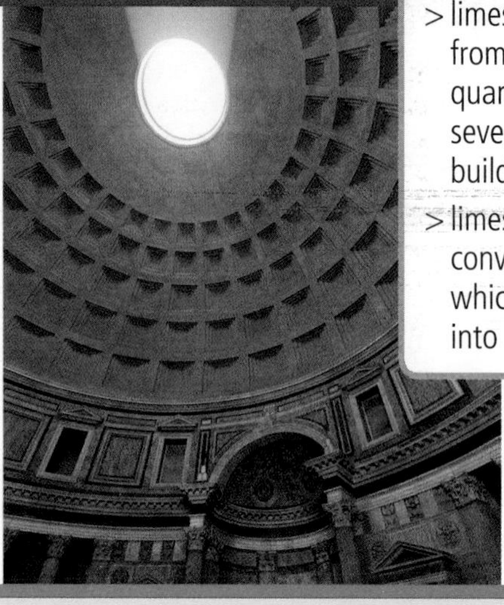

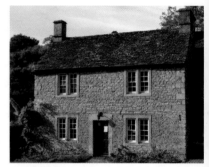

**FIGURE 1**: The Pantheon – a 2000-year-old concrete dome.

### Essential notes

> limestone is obtained from the ground by quarrying and has several uses in the building industry

> limestone can be converted into quicklime, which can be converted into slaked lime

## Obtaining and using limestone

### Obtaining limestone

**Limestone** is a sedimentary rock. It started life many millions of years ago as shells and skeletons of sea creatures and coral. It is mainly **calcium carbonate**, $CaCO_3$.

Limestone is quarried from the ground. The top layer of soil is removed and explosives are used to blast the rock into smaller pieces of limestone.

Quarries have to be situated where the rock occurs. This might even be in a National Park. However, there are strict rules nowadays. New quarries or extensions to quarries must be approved by the local planning authirity.

A **quarry** can bring both advantages and disadvantages to an area.

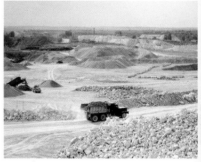

**FIGURE 2**: Buildings constructed from limestone are found in the UK and around the world.

**For**

> Quarries provide jobs and support the local economy, for example, better roads and healthcare facilities.

> Quarries must meet public demand for products made using limestone.

> Some old quarries have been restored and are now important wildlife sites or provide leisure facilities.

**Against**

> Quarries are large unsightly scars on the landscape.

> Traffic may increase, with congestion and vibrations from heavy lorries.

> There may be damage to wildlife habitats as a result of loss of land, noise and dust pollution.

**FIGURE 3**: There are arguments for and against quarrying limestone. What do you think?

### Using limestone

Limestone is an important building material and a raw material from which various other products such as **quicklime**, **cement**, **mortar**, **concrete** and **glass** are made. It is also used in the extraction of **iron** from its ores.

### Did you know?

Limestone, chalk, marble and egg shells are made of the same chemical compound – calcium carbonate.

### QUESTIONS

1 How is limestone obtained?

2 Explain why limestone is an important raw material.



Q ... limestone

## Limestone, quicklime, slaked lime and concrete

### Limestone to quicklime

When limestone (calcium carbonate) is heated, it thermally decomposes to give quicklime (calcium oxide) and carbon dioxide.

limestone ➜ quicklime + carbon dioxide

or, using their chemical names:

calcium carbonate ➜ calcium oxide + carbon dioxide

$$CaCO_3(s) \rightarrow CaO(s) + CO_2(g)$$

Quicklime has been used in self-heating cans of coffee and soup – there is no need for a fire or a stove. The cans contain calcium oxide and water. When the two are mixed energy is released and heats the coffee or soup. Of course, the coffee or soup must not be mixed with the calcium oxide!

### Quicklime to slaked lime

Adding water to quicklime makes slaked lime (calcium hydroxide). A great deal of heat is given out as the reaction happens.

quicklime + water ➜ slaked lime

or, using their chemical names:

calcium oxide + water ➜ calcium hydroxide

$$CaO(s) + H_2O(\ell) \rightarrow Ca(OH)_2(s)$$

Slaked lime is an alkali. Farmers and gardeners spread it on soil that is too acidic. Sometimes it is added to lakes that have become acidic as a result of acid rain.

It was used, and still is in some parts of the world, to make whitewash and lime plaster.

### Cement

Cement is made by heating limestone and clay in a kiln (a very hot oven). It is a fine powder that sets rock hard when mixed with water. It is not often used directly in construction work. Usually it is used to make mortar and concrete.

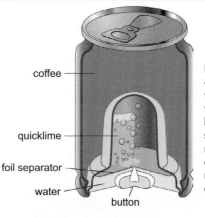

coffee

quicklime

foil separator

water

button

**FIGURE 4**: Hot coffee made in the can. Press the button to break the foil separator and mix water with quicklime. In minutes your coffee is ready.

**FIGURE 5**: Making slaked lime from quicklime. Just add water – carefully!

### QUESTIONS

**3** Give the chemical names and formulae for (a) quicklime (b) slaked lime.

**4** Describe how slaked lime is obtained from limestone.

**5** Describe how cement is obtained from limestone.

## Acid attack

Limestone is weathered by rainwater. Rainwater is slightly acidic owing to dissolved carbon dioxide that has formed carbonic acid, $H_2CO_3$

$$H_2O(\ell) + CO_2(g) \rightarrow H_2CO_3(aq)$$

Carbonic acid then reacts with limestone, dissolving it to give a solution of calcium hydrogencarbonate.

$$H_2CO_3(aq) + CaCO_3(s) \rightarrow Ca(HCO_3)_2(aq)$$

The damage is even greater with acid rain (rain made more acidic because of pollutants such as **sulfur dioxide** produced by burning **fossil fuels**).

### QUESTIONS

**6** Explain why limestone is eroded more quickly by acid rain.

🔍 ... slaked lime ... acid rain

# Materials from limestone

## Essential notes

> limestone is used in the manufacture of quicklime, cement and glass

> mortar and concrete are made from cement and aggregate – sand in mortar, and sand and gravel in concrete

**THE SCIENCE IN CONTEXT** Manufacturing cement, mortar, concrete and glass from limestone involves some chemical reactions that are complex and still not fully understood. Materials scientists study the structures and properties of these materials to make sure that they are used effectively.

## Crystal Palace

The Crystal Palace is constructed from glass and cast iron. It was assembled in sections brought to the site. Limestone is used when making both glass and iron.

FIGURE 1: The Crystal Palace was more than 560 metres long, 120 metres wide and 33 metres high. 84 000 square metres of glass was used in its construction.

## Limestone as a raw material

Limestone is a much-used building stone. It is also used to make four closely related materials used by the construction industry – cement, mortar, concrete and glass.

> Cement is made from limestone and clay in a kiln. However, it is rarely used alone. It is used to make mortar and concrete.

> Mortar is used to bind bricks, stones and other building materials. It is made by mixing cement, sand and water. Originally slaked lime was used instead of cement. This lime mortar is still used occasionally to restore old buildings.

> Concrete is made by mixing cement, sand, gravel (or crushed stone) and water. The composition of concrete varies, but cement is usually 10–15% by mass of the concrete mix. Concrete is a very strong building material, but it is also **brittle**. So it is often combined with **steel** rods to make steel reinforced concrete.

FIGURE 2: Cement, sand and water are mixed to make mortar.

> Glass is an important construction material because:
> > it is transparent
> > it is rigid (though with some flexibility)
> > is can be made in large sheets
> > it is a good thermal **conductor**
> > it is weather-resistant.

On the down side it is fragile and breaks easily, though modern technologies enable glass to be strengthened.

There are different types of glass. Soda-lime glass is the most common and accounts for 90% of all glass produced. It is made by heating a mixture of sand, soda ash and limestone.

## QUESTIONS

1 Describe the difference between mortar and concrete.

2 What is soda-lime glass made from?

3 Suggest why glass is a popular choice for the outside of buildings.

FIGURE 3: The unusual shape of City Hall, offices of the Mayor of London, is intended to reduce the building's surface area and thus improve energy efficiency.

# Manufacture

## Manufacture of quicklime and cement

Quicklime is manufactured in kilns. Crushed limestone is heated to above 900 °C in a kiln. It is cooled by a flow of cold air.

$$CaCO_3(s) \rightarrow CaO(s) + CO_2(g)$$

Cement is also made in a kiln. Limestone and clay are crushed and heated in a rotating kiln at temperatures of up to 1450 °C. Often sand and iron oxide are also put into the mix. Chemical reactions take place. Carbon dioxide and water are produced and driven off. New calcium silicate and calcium aluminate compounds are formed. The cement is cooled, finely ground and stored in a dry place.

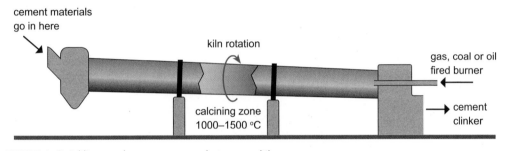

cement materials go in here

kiln rotation

gas, coal or oil fired burner

calcining zone 1000–1500 °C

cement clinker

**FIGURE 4**: Quicklime and cement are made in rotary kilns. Why are quicklime and cement stored in dry places?

## Manufacture of glass

The chemical compounds in the raw materials used to make glass are:

> sand (silicon dioxide, $SiO_2$)

> soda ash (sodium carbonate, $Na_2CO_3$)

> limestone (calcium carbonate, $CaCO_3$)

These are mixed with recycled glass and fed into a glass-melting furnace. Furnaces operate at 1600 °C, 24 hours a day, seven days a week. Most are fired with **natural gas**. A furnace **melting** 300 000 kilograms per day uses about 32 000 cubic metres of natural gas each day.

### Did you know?

Each year, the UK glass industry makes about 2.8 million tonnes of glass. This sells for about £1500 million.

### QUESTIONS

**4** Name the two raw materials needed to make cement.

**5** Atoms of which elements combine to make (a) sand (b) soda ash (c) limestone?

# Structure of silicon dioxide and glass

Silicon dioxide is a crystalline solid. It has a high melting point, it is hard and does not conduct electricity. This is because silicon dioxide has a giant structure, with silicon and oxygen atoms held together by covalent bonds.

The reactions between silicon dioxide, sodium carbonate and calcium carbonate break down the crystalline structure of silicon dioxide, leaving networks of silicon and oxygen atoms. This is why glass has different properties from sand. Some oxygen atoms are left with a negative **charge**. These negative charges are balanced by positively charged sodium ions and calcium ions.

**FIGURE 5**: Quartz crystals. Quartz is silicon dioxide.

### QUESTIONS

**6** The chemistry of glass-making is complex. Find out more about the networks of silicon and oxygen atoms in glass.

# Metals

**THE SCIENCE IN CONTEXT** Metals are a traditional construction material. Their structures, physical properties and chemical reactivity are investigated by chemists, materials scientists and metallurgists. On the whole, metals have similar properties, but there are important differences as well. The construction industry makes use of these differences and picks the most suitable metal for a particular job.

## Essential notes

> metals have characteristic properties that make them useful for construction work: malleable, strong, hard, good electrical conduction and resistance to corrosion

> metals such as iron, steel, aluminium, lead and copper are used in the construction industry

> how metals are used in the construction industry depends on their properties

## Steel bridges

Many people think that Sydney Harbour Bridge in Australia was based on the Tyne Bridge in Newcastle. However, although construction of the Tyne Bridge was finished in 1929 and the Sydney Harbour Bridge in 1932, work started later on the Tyne Bridge. Both were made of steel.

**FIGURE 1**: Tyne Bridge in Newcastle (top) and Sydney Harbour Bridge (bottom). Both have arches from which the bridge platform is suspended on steel cables.

## Properties of metals

In general, metals are:

> good conductors of heat and electricity

> malleable (can be pressed into sheets)

> strong

> tough (not brittle)

> hard (do not dent easily).

Some metals corrode (react with air, rainwater, seawater and acid rain) while others are resistant to corrosion.

### Metals used in construction

The metals most often used are **cast iron**, steel, aluminium, copper and lead.

| Metal | Uses |
|---|---|
| cast iron and steel | beams, girders, frameworks for buildings, bridges and reinforcing concrete |
| aluminium | window frames, doors, cladding, frameworks for light-weight buildings such as conservatories |
| copper | electrical wiring and cables, pipes for plumbing, hot water cylinders, roofing |
| lead | flashing (seals between roofs and vertical walls) |

### Did you know?

The Sydney Harbour Bridge can rise or fall up to 18 cm. This is because the cables are made of steel which, like all metals, expands when heated and contracts when cooled.

## QUESTIONS

**1** What does 'malleable' mean?

**2** Why is the chemical reactivity of metals important when considering their use in construction work?

**3** Which metals are most commonly used in the construction industry?

Q ... gcse metal properties and uses

# Cast iron, steel and other metals

## Cast iron and steel

Cast iron is used in construction. It is 95% iron, with at least 2% carbon and some silicon. It is strong and corrosion resistant. However, it is also brittle.

Steel is an **alloy** of iron and carbon (about 0.1 to 2%). There many different types of steel depending on which other metals are are also added to the iron. Steels are less brittle than cast iron, but they still corrode in moist atmospheres, apart from **stainless steel** (which contains 11% chromium).

## Other metals

Aluminium, copper and lead are also used in construction.

| Metal | Properties that make it useful in construction | |
|-------|-----------------------------------------------|--|
| aluminium | strong, malleable, light (has a low density), corrosion resistant | |
| copper | malleable, corrosion resistant, good conductor of electricity | |
| lead | malleable, very resistant to corrosion | |

FIGURE 2: Construction of the Iron Bridge in Shropshire was completed in 1779 after four years' work. It is one of the first structures to be made almost entirely from cast iron.

## QUESTIONS

**4** What properties of aluminium make it suitable to use in a greenhouse?

**5** Why is electrical wiring made of copper?

**6** Lead is used to make a waterproof joint between a roof covering and vertical surfaces such as walls. What properties make it good for this?

# Reactivity series

Corrosion is a chemical reaction that happens when a metal comes in contact with oxygen and water. The reactivity series list metals according to how reactive they are. The most reactive metals appear at the top of the list. Carbon and hydrogen are not metals, but they provide reference points in the series.

Three of the metals used in the construction industry are below carbon in the reactivity series. Aluminium is quite high in the reactivity series yet very resistant to corrosion. This is because aluminium is reactive and forms a surface coating of aluminium oxide on contact with air. However, once formed it protects the aluminium from further reaction.

TABLE 1: The reactivity series.

| potassium | K |
|-----------|---|
| sodium | Na |
| calcium | Ca |
| magnesium | Mg |
| aluminium | Al |
| zinc | Zn |
| carbon | C |
| Iron | Fe |
| lead | Pb |
| hydrogen | H |
| copper | Cu |
| silver | Ag |
| gold | Au |

Increasing reactivity ↑

## QUESTIONS

**7** Which three metals used in the construction industry are below carbon in the reactivity series?

**8** Which two unreactive metals are not used in the construction industry? Explain why they are not.

# Polymers

**THE SCIENCE IN CONTEXT** Polymers are construction materials that have been manufactured. They have partly replaced wood, a natural construction material, and metals in a number of applications such as window and door frames. Polymers are made by chemical reactions which scientists can control so that the properties of a polymer are modified to increase their suitability for a particular use.

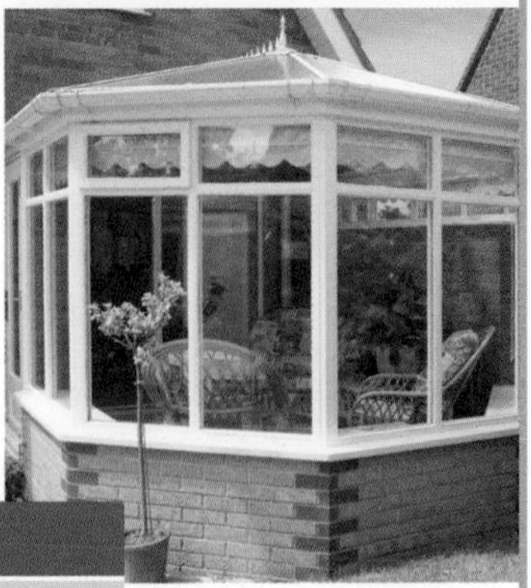

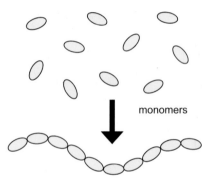

FIGURE 1: Which parts of this conservatory are made from polymers?

## Plastic houses

A company in Swansea builds houses using panels made of recycled waste plastic – 18 000 kg for each house. The panels are fireproof, waterproof and stronger than concrete.

## Polymers

### Making polymers

**Polymers** (often called **plastics**) are made using chemicals obtained from crude oil. Small molecules join together to form very large molecules. Small molecules that can join together are **monomers**. The very large molecules they form are **polymers**.

Two examples are:

> polyethene (popularly known as polythene) – made from ethene, $C_2H_4$

> polypropene (also called polypropylene) – made from propene, $C_3H_6$

### Properties and uses of polymers

Different polymers are used in the construction industry. In general, polymers are:

> flexible

> poor conductors of heat and electricity

> waterproof

> corrosion resistant.

Most polymers have low melting points. This means that can be melted and poured into moulds to make items of required shapes.

| Polymer | Uses |
|---|---|
| polyethene ('polythene') | damp-proof membranes, coatings |
| polypropene | sound insulation, water pipes, waste pipes |
| poly(2-methylpropene) | adhesives and glazing sealant, waterproof membranes |
| polyvinyl chloride (PVC) | cladding, pipes, guttering, wiring and cable insulation, window frames, doors |
| polystyrene | thermal (heat) and sound insulation |
| polycarbonate | glazing, conservatory roofs |

monomers

part of the polymer

FIGURE 2: Many thousands of small molecules join together to make a large polymer molecule.

## Remember

The chemical reaction from which a polymer is made is called polymerisation.

## Did you know?

The first use of polyethene was during World War II to insulate radar cables. At the time it was a military secret.

Q ... polymers

FIGURE 3: The concrete base for this house is covered with a plastic damp-proof membrane before more concrete is added to bring the floor to the required level.

## Polymerisation of ethene

Ethene is a gas obtained by processing crude oil by a process called **cracking**. Ethene is an **alkene**. It has the formula $C_2H_4$. All alkenes have a carbon–carbon double bond, C=C. It is this double bond that enables them to join together to make polymers.

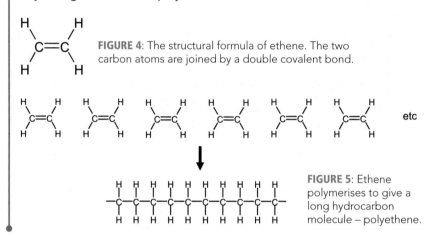

FIGURE 4: The structural formula of ethene. The two carbon atoms are joined by a double covalent bond.

etc

FIGURE 5: Ethene polymerises to give a long hydrocarbon molecule – polyethene.

## Types of polymers

People often talk about plastics rather than polymers, but there is a difference. Broadly there are two types of polymer: **thermoplastic** and **thermosetting**.

Thermoplastic polymers can be softened or melted and shaped. This process can repeated over and over again. The polymer is a long chain of atoms. Polyethene, polypropene, poly(2-methylpropene, polyvinyl chloride (PVC), polystyrene and polycarbonate are all examples.

Thermosetting polymers change chemically when they are heated. But it is a one-way process – they cannot be reheated and re-shaped. The polymer is a network of atoms rather than a chain. Two examples are:

> urea-formaldehyde resin, used in medium-density fibreboard (MDF)

> epoxy resin, used in glass-reinforced plastic.

# Ceramics

**THE SCIENCE IN CONTEXT** Look at the roofs of most buildings and you will be looking at ceramic materials. Ceramic materials have strengths and weaknesses. They have excellent resistance to chemicals, are very poor conductors and are very hard. But they break easily. Understanding these properties and the structures of ceramics enables the building industry to put them to good use.

### It's just boring

The Channel Tunnel was built using giant drills. They bored into the rock, grinding it up so that it could be removed from the tunnel. The tips of the drills were made of a ceramic material called tungsten carbide.

**FIGURE 1**: Tungsten carbide was used for the tips of the Channel Tunnel drills because of its hardness and resistance to wear.

## Properties and uses of ceramics

### Properties of ceramics

Ceramics are solid inorganic substances that form only at very high temperatures.

Each **atom** in a ceramic is bonded to several others around it in a giant lattice (a giant structure). Because of this, ceramics have very high melting points and are:

> hard
> brittle and non-flexible
> corrosion resistant
> resistant to chemicals such as acids, alkalis and bleach
> very poor conductors of heat and electricity (in other words, good insulators).

### Uses of ceramics

Ceramics have several uses in and around the home:

| Purpose | Examples |
|---|---|
| construction and decoration | bricks, roof tiles, floor and wall tiles, drainage pipes, chimney lining |
| pottery products | bathroom basins and toilets |
| specialist industrial materials | lining for furnaces, insulators on power transmission lines |

**FIGURE 2**: Clay roof tiles are an example of ceramics.

### Did you know?

Diamond was once the hardest material known. However, a material called wurtzite boron nitride has been made. It is 58% harder than diamond.

## QUESTIONS

1 What type of structure do ceramics have?

2 Ceramics are hard and brittle. Explain the difference in these two properties.

3 Name two purposes for which ceramics are used.

Q ... gcse ceramics

# Bricks, clay, china and porcelain

Modern bricks are made by pressing clay into moulds and heating them in a kiln at about 1000-1200 °C. All the water is driven off and new chemical substances are formed. The biggest use for bricks is for building houses.

Similarly, tiles for roofs, floors and walls can be made from clay. They last a long time and are probably the most durable (hard-wearing and long-lasting) option. Roof tiles are resistant to weathering, but quite heavy and brittle (walking on them can break them). Floor and wall tiles are easy to clean and are not attacked by chemicals used, for example, in the kitchen and bathroom.

Various materials are used to make bathroom basins and toilets. However, vitreous china (a ceramic) is the most common. It is made by a pumping a mixture of clay, crushed glass and some other finely ground minerals into a mould. The mixture is fired at around 1200 °C in a kiln.

None of these applications take advantage of the insulating properties of ceramics. However, the use of porcelain for insulating power transmission lines does.

FIGURE 3: This bathroom wash basin and toilet are made from vitreous china. What else is made from ceramics?

FIGURE 4: The white caps are porcelain electrical insulators.

## QUESTIONS

**4** Clay roof tiles are an example of ceramics. Which properties of ceramics (a) make them suitable to use (b) are disadvantageous?

**5** Why are white porcelain caps used in power transmission lines?

# The nature of ceramic materials

Traditional ceramics include clay products such as pottery, bricks and tiles. Advanced ceramics include carbides, oxides and nitrides.

All ceramics have giant structures. This explains why they are hard but brittle. The lattice does not let particles move so ceramics are difficult to dent or scratch, and are not flexible. Applying sufficient force just breaks the lattice apart.

Ceramics have strong bonds between atoms. Aluminium oxide, $Al_2O_3$, has ionic bonds. Silicon carbide (SiC) has covalent bonds. That is why ceramics have very high melting points. It takes a lot of energy to break the bonds so that particles can separate and move around freely as a liquid. Electrons are fixed in these strong bonds. They are not free to move, so no current can flow. This is why ceramics are good electrical insulators.

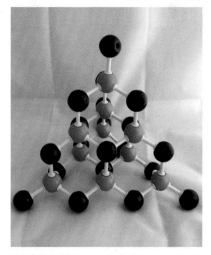

FIGURE 5: Silicon carbide is a ceramic. It has the same structure as diamond. In this model, green spheres represent silicon atoms and black spheres represent carbon atoms.

## QUESTIONS

**6** Describe the structure of silicon carbide and explain why it has a very high melting point, is hard, brittle and a good electrical insulator.

# Composites

**THE SCIENCE IN CONTEXT** Composites have been around for a very long time. Yet they are at the cutting edge of materials research and development. New polymers are being combined with metals or ceramics. New ceramics are being combined with metals. Metal foams (a composite of metal and air) are finding valuable applications. They may be the future of construction materials.

## Mud houses

Mud bricks were first used about 6000 years ago. They are strong under compression, but break easily if bent. Straw, on the other hand, has good tensile strength, but is very weak when crumpled. Incorporating straw into mud bricks gave the best of both worlds – composite bricks that made excellent building materials.

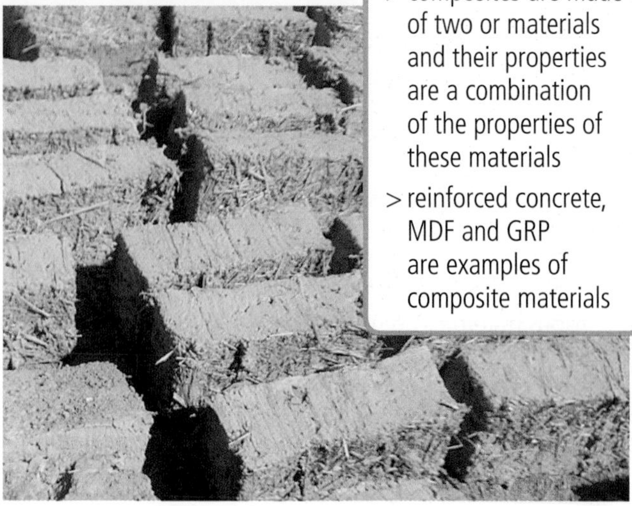

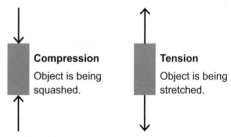

FIGURE 1: Adobe bricks are made from mud (clay and water), sand and straw.

## Combining materials

Sometimes, metals, polymers and ceramics are not suitable for a particular purpose. For the right mix of properties you need a **composite**. Composites are made of two or more materials. Their properties are a combination of the properties of these materials.

Mixing straw and mud to make bricks is thousands of years old. It combined the tensile strength of straw with the compressive strength of dried mud. This idea has been mimicked in lots of composite materials.

Examples of composite materials used in construction and for other purposes include:

> reinforced concrete – a composite of concrete and a metal (usually steel)

> MDF (medium density fibre board) – a composite of wood fibres and a thermosetting polymer

> GRP (glass-reinforced-plastic) – a composite of glass fibres and a thermosetting polymer

### Strength

Strength is an important property of materials used in construction. There are two types of strength:

> tensile strength – resistance to being stretched when each end is pulled in opposite directions

> compressive strength – resistance to being squashed when each side is being pushed together.

Some materials are strong in **compression**, but weak in tension. Other materials are the opposite.

## Did you know?

Adobe buildings are among the oldest buildings still standing.

**Compression**
Object is being squashed.

**Tension**
Object is being stretched.

FIGURE 2: The difference between compressive strength and tensile strength.

## QUESTIONS

1 Explain the difference between compressive strength and tensile strength.

2 In straw and mud bricks which component provides
(a) compressive strength
(b) tensile strength?

3 Explain why MDF and GRP are composites.

## Examples of composites

### Reinforced concrete

Concrete has good compressive strength (which is why very heavy structures can stand on concrete foundations), but poor tensile strength. Metals have poor compressive strength (that is why they are malleable), but good tensile strength.

Embedding metal rods, wires, mesh or cables in concrete gives you the best of both worlds — reinforced concrete.

### Medium density fibre board (MDF)

MDF is flat, stiff, has no knots and is easy to saw and sand down. It can be used instead of plywood or chipboard (which are also composites).

MDF is a composite of wood fibres and a thermosetting polymer. The fibres have tensile strength and the thermosetting polymer has compressive strength. MDF has both these properties.

### Glass-reinforced-plastic (GRP)

GRP (sometimes called fibreglass) was the first modern composite. It consists of two materials – glass fibres and a thermosetting polymer. It combines the tensile strength of glass fibres with the compressive strength of the thermosetting polymer.

GRP is a resilient material. After bending, twisting, stretching or squashing, it returns to its original shape. It is also light and has a very good strength to weight ratio.

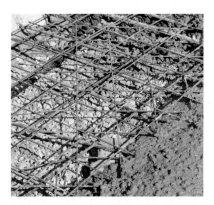

**FIGURE 3**: Concrete is poured over metal mesh to make a reinforced concrete floor.

**FIGURE 4**: Why is this garage door made from GRP?

## QUESTIONS

**4** What do reinforced concrete, MDF and GRP have in common?

**5** How does reinforcing concrete with metal improve its properties as a construction material?

**6** GRP is a resilient material. What does this mean?

## Advanced composites

MDF and GRP are well-established composites with fibres embedded in a polymer. Recently, two types of fibres (carbon and *Kevlar*®) have been used with a thermosetting polymer to make fibre-reinforced plastic – an example of an advanced composite.

The advantages of these carbon and *Kevlar*® fibres over glass fibres are:

> greater tensile strength and stiffness

> lower density.

They are expensive. The main user is the aerospace industry. The construction industry is interested in using them to strengthen existing structures or as an alternative reinforcing material to steel.

## QUESTIONS

**7** Explain the purpose of the two components in carbon-fibre-reinforced plastic and compare its properties with those of glass-reinforced plastics.

# Preparing for assessment: Processing and analysing data

To achieve a good grade in science, you not only have to know and understand scientific ideas, but you need to be able to apply them to other situations and investigations.

Connections: This task relates to Unit 2 context 3.4.2.1 Materials used to construct our homes. Working through this task will help you to prepare for the practical investigation that you need to carry out for your Controlled Assessment.

## ✳ Investigating the strength of concrete

Freddie and Laura wanted to investigate how the thickness of concrete affects its strength. They knew that other people would have already carried out some work on this topic. Before planning their investigation, they decided to carry out research to find different methods that they could use.

Freddie searched the internet and found this method:

**Method A** A steel ball is dropped onto a concrete block. Each time it is dropped from 10 cm higher than the previous height, until the concrete block cracks. This height then indicates the strength of the block.

steel ball

concrete block

Laura looked through a science textbook and found this alternative method:

**Method B** A concrete block is placed on two supports on the bench. By placing weights on top of the unsupported part of the block, as shown, it is possible to measure the force in newtons needed to crack the block. This force is a measure of the strength of the block.

masses

concrete beam

supports

### Investigation carried out by Laura and Freddie

Freddie and Laura made three sets of concrete blocks. Each set had a range of thicknesses, from 1.0 cm thick to 3.0 cm thick.

Two sets of concrete blocks were investigated using Method A (taken from the internet), one by Freddie and the other by Laura. Here are their results.

### Freddie's results (using Method A)
1 cm thick: 20 cm high;   1.5 cm thick: 30 cm;   2 cm thick: 50 cm;   2.5 cm thick: 80 cm;   3 cm thick: 90 cm

### Laura's results (using Method A)
1 cm thick: 20 cm high;   1.5 cm thick: 30 cm;   2 cm thick: 50 cm;   2.5 cm thick: 60 cm;   3 cm thick: 90 cm

The third set of concrete blocks was tested using Method B (taken from a science textbook) by Freddie and Laura working together. Here are their results.

1 cm thick: 40 N;   1.5 cm thick: 60 N;   2 cm thick: 80 N;   2.5 cm thick: 100 N;   3 cm thick: 120 N

## ✴ Processing data

**1.** Construct and complete a results table for the data collected by Laura and Freddie using Method A (taken from the internet).

**2.** Construct and complete a results table for Freddie's data and Laura's data collected using Method B ( taken from a textbook).

Make sure that the headings of the rows and columns are complete, and that you state the units. You also need to be consistent in terms of the appropriate number of significant figures. For example, thicknesses should all be given to two significant figures, so 1.0 cm rather than 1 cm.

Again, make sure that the headings of rows and columns are complete and units are given.

## ✴ Analysing data

**3.** Use your table of data collected from Method A to draw graphs of Freddie's results and of Laura's results.

**4.** Look at the data points of both graphs. Identify any anomalous results and explain why these measurement(s) should be ignored or repeated.

**5.** Use your table of data collected from Method B to draw a graph of Freddie's and Laura's results.

**6.** Look at the two graphs and describe the trend in each case.

**7.** Using the graph of data collected from method B, calculate the gradient (in the units newtons per centimetre, N/cm).

**8.** Use this graph to predict the force needed to break a 2.25 cm thick concrete block.

Remember to choose a suitable scale for your graph and to label the axes. The independent variable (the thickness of the concrete block) should be placed on the x-axis.

You could plot the two sets of data on the same graph or use two separate graphs. You need to decide which is easiest for spotting trends and patterns in the data.

The line should be a straight line or a curve, whichever you think best fits the data. Remember that the (graph) line must pass through the origin (0,0).

Anomalous results are those that do not appear to fit the pattern of the other results. Sometimes the measurement can be repeated (if it is non-destructive), but sometimes it cannot if it is destructive (for example, if the concrete block is broken in the measurement).

Again, the independent variable should be placed on the x-axis, and a line drawn through the points. The line should be a straight line or a curve, whichever you think best fits the data. Remember that the line must pass through the origin (0,0).

Try to use terms such as directly proportional if the line is straight. A curve indicates that the relationship between the independent and dependent variable is more complex.

Remember that a gradient is calculated from the change in y-axis divided by the change in the x-axis.

You can do this directly from the graph or by using the gradient as a conversion factor.

# Fuels

**THE SCIENCE IN CONTEXT** It is not easy to imagine life without fuels. We use them to do so much. Most fuels are hydrocarbons; organic compounds made from carbon and hydrogen atoms. They are obtained from crude oil. Chemists have studied the physical properties, chemical structure and chemical reactivity of hydrocarbons in great detail.

## Cooking on gas

In recent years, gas barbecues have become popular. The gas is usually butane, obtained by processing crude oil. It is liquid because of the high pressure at which it is stored, but it is still called 'bottled gas'.

FIGURE 1: This barbecue is fuelled by butane. You can see the 'bottled gas' beneath the grill.

### Essential notes

> suitable fuels for cooking and heating our homes and for providing transport include natural gas, bottled gas, petrol, diesel, kerosene and heating oil

> hydrocarbons are made from carbon and hydrogen atoms only

> alkanes (one type of hydrocarbon) have the general formula $C_nH_{2n+2}$

## Why do we need fuel?

Fuels are used to cook, heat homes and to power cars, ships, aeroplanes, lorries and buses. A **fuel** is a substance that burns in air, releasing energy stored in it.

The vast majority of fuels used today are **fossil fuels** – natural gas, crude oil and **coal**. Before fossil fuels, people used wood and animal fats. Even today, many homes have wood-burning fires and stoves. Unlike fossil fuels, wood can be replaced – you just plant more trees.

**Natural gas** is mainly **methane**, $CH_4$. It is used for gas cookers, gas fires and gas central heating systems. **Fractions** distilled from **crude oil** are used for cooking, heating and transport.

### Hydrocarbons

Natural gas and the compounds in crude oil are **hydrocarbons**. Hydrocarbons are made from carbon and hydrogen atoms only. When they burn, the hydrocarbons react with oxygen in the air. Carbon dioxide and water are products of the reaction.

hydrocarbon + oxygen ➞ carbon dioxide + water

This is a **combustion reaction**. The reactions release lots of heat, making them good fuels.

| Fraction | Use |
|----------|-----|
| petrol | fuel for cars |
| kerosene | aircraft fuel, paraffin lamps and heaters |
| diesel oil | fuel for cars, lorries and buses |
| fuel oil | heating oil and fuel for ships |

FIGURE 2: A paraffin heater makes sure the temperature in the greenhouse does not fall too low.

## Did you know?

Cavemen used coal for heating and cooking.

## QUESTIONS

1 What is the main chemical compound in natural gas?

2 Which fractions of crude oil can be used as fuel for cars?

3 Explain what a combustion reaction is.

# Alkanes

**Alkanes** are hydrocarbons that have only single covalent bonds in their molecules. The four simplest alkanes are methane, ethane, propane and butane.

From the alkanes in Figure 3 you can probably see a pattern emerging. As you move from one alkane to the next in the series one –CH$_2$– unit has been added.

The general formula for an alkane is C$_n$H$_{2n+2}$

Octane has eight carbon atoms. Therefore it will have $(2 \times 8) + 2 = 18$ hydrogen atoms.

## Combustion of alkanes

Using the chemical formulae of the reactants and products, the word equation is:

methane + oxygen → carbon dioxide + water

Similarly, the word equations for the **combustion** of other alkanes may be written:

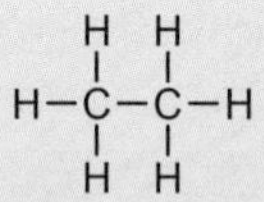

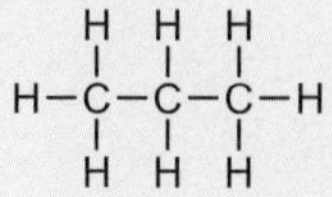

> ethane + oxygen → carbon dioxide and water

> propane + oxygen → carbon dioxide and water

> butane + oxygen → carbon dioxide and water

| Alkane | Molecular formula: shows the number of atoms in a molecule | Structural formula: shows the bonds between atoms in a molecule |
|---|---|---|
| methane | CH$_4$ | H–C–H (with H above and below) |
| ethane | C$_2$H$_6$ | H–C–C–H (with H above and below each C) |
| propane | C$_3$H$_8$ | H–C–C–C–H (with H above and below each C) |
| butane | C$_4$H$_{10}$ | H–C–C–C–C–H (with H above and below each C) |

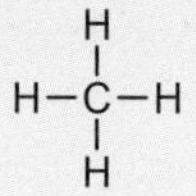

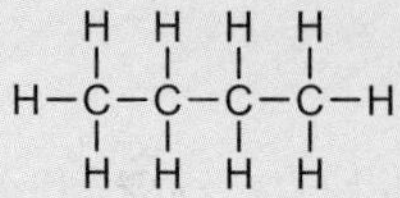

FIGURE 3: Molecular and structural formulae of alkanes.

## QUESTIONS

**4** Alkanes are made from the atoms of which elements?

**5** Octane has eight carbon atoms. What is its molecular formula?

**6** Sketch the structural formulae of (a) pentane, C$_5$H$_{12}$ (b) hexane, C$_6$H$_{14}$.

# Formulae of alkanes

There are several factors that affect how a molecule reacts, including:

> the types and numbers of atoms that it is made from

> the way in which the atoms have bonded to one another

> its shape.

A **molecular formula** tells you about the numbers and types of atoms.

A **structural formula** tells you about the numbers and types of atoms and which are bonded to which.

A **displayed formula** shows the shape of the molecule.

Examples of molecular and structural formulae are shown in Figure 3. Figure 4 shows the displayed formulae of these alkanes.

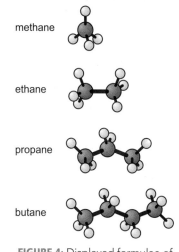

methane

ethane

propane

butane

FIGURE 4: Displayed formulae of the first four of the alkane series.

## QUESTIONS

**7** Give reasons for having different types of chemical formulae.

# Burning fuels

**THE SCIENCE IN CONTEXT** Like all chemical substances, hydrocarbons store energy. Some of this energy is released when hydrocarbons burn and is transferred to the surroundings, which become hotter. This is why they are useful as fuels. Scientists, technologists and engineers combine their knowledge and skills to make most efficient use of the energy released for cooking, heating and transport.

FIGURE 1: The search for new reserves of gas and oil goes on.

## Finite fuels

Supplies of fossil fuels are finite – eventually they will run out. It is hard to say when, but it may be about 50 years for crude oil and natural gas and a few hundred years for coal.

### Essential notes

> fossil fuels are a finite, non-renewable source of energy

> burning fossil fuels produces carbon dioxide (linked to global warming) and pollutants, such as sulfur dioxide

> the amount of heat released when a hydrocarbon combusts is proportional to the number of carbon atoms in the hydrocarbon molecule

> the combustion of hydrocarbons can be shown using balanced symbol equations

## Problems with fuels

Fossil fuels are a **finite, non-renewable** source of energy. *"What do we do when the fossil fuels run out?"* is a very real question that scientists, technologists and engineers need to answer. There are other problems, too.

### Global warming

Fossil fuels contain hydrocarbons. They produce carbon dioxide when burned.

hydrocarbons + oxygen → carbon dioxide + water

Carbon dioxide in the air helps to sustain life. It traps heat from the Sun. This helps to keep Earth's surface warm. Burning fossil fuels increases the amount of carbon dioxide in the air. More heat is trapped and the air and Earth's surface warm up. This effect is called **global warming**.

### Pollution

Burning fossil fuels produces **pollutants**.

Often the hydrocarbons do not combust completely and produce smoke and carbon monoxide (a poisonous gas).

Fossil fuels contain impurities. When the fossil fuel burns these react with oxygen, producing pollutants such as sulfur dioxide. Sulfur dioxide oxidises in air to sulfur trioxide. This dissolves in water, in clouds, to make sulfuric acid.

sulfur trioxide + water → sulfuric acid

These clouds produce **acid rain**. While sulfur dioxide is the major cause of acid rain, oxides of nitrogen are another.

FIGURE 2: Why are carbon monoxide detectors often fitted near to gas-fired water heaters in the home?

FIGURE 3: The devastating effect of acid rain on a pine forest.

## QUESTIONS

1 What is a **'finite resource'**?

2 What are greenhouse gases and how are they linked to global warming?

3 Draw a flow chart to show how acid rain forms.

Q ... gcse burning fuels

# Energy from burning hydrocarbons

The energy released when a hydrocarbon burns is called the **heat of combustion**.

For the moment do not worry about the units – it is the relative size of the heat of combustion values that is important.

Looking at the structural formulae of alkanes, you will see that a $-CH_2-$ unit is added each time:

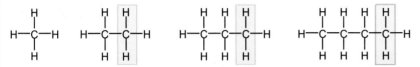

The heat of combustion of an alkane is proportional to the number of carbon atoms in its molecules. This shows that each $-CH_2-$ unit produces a fixed amount of energy.

| Alkane | Heat of combustion (kJ/mol) |
|--------|------------------------------|
| methane | 890 |
| ethane | 1560 |
| propane | 2219 |
| butane | 2877 |
| pentane | 3509 |
| hexane | 4163 |

**TABLE 1**: Heats of combustion of alkanes.

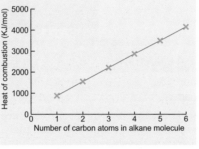

**FIGURE 4**: Heats of combustion of alkanes.

## Did you know?

Forty litres of petrol (enough to fill the tank of a small car) produces over 90 kilograms of carbon dioxide.

## QUESTIONS

**4** Describe how the heat of combustion of an alkane depends on the number of carbon atoms in its molecules.

**5** Suggest a reason for the pattern you see.

# Combustion equations (Higher tier)

When natural gas burns, methane reacts with oxygen to release the energy needed. The **balanced equation** for this is:

$$CH_4 + 2O_2 \rightarrow CO_2 + 2H_2O$$

Balanced equations can be written for the combustion of all alkanes. For example:

> ethane      $2C_2H_6 + 7O_2 \rightarrow 4CO_2 + 6H_2O$

> propane      $C_3H_8 + 5O_2 \rightarrow 3CO_2 + 4H_2O$

> butane      $2C_4H_{10} + 13O_2 \rightarrow 8CO_2 + 10H_2O$

> pentane      $C_5H_{12} + 8O_2 \rightarrow 5CO_2 + 6H_2O$

A pattern emerges (see table 2).

The general equation for the combustion of an alkane is:

$$C_nH_{2n+2} + (1.5n + 0.5)O_2 \rightarrow nCO_2 + (n + 1)H_2O$$

| Alkane | Ratio of alkane molecules to oxygen molecules |
|--------|------------------------------------------------|
| methane | 1 $CH_4$ to 2 $O_2$ |
| ethane | 1 $C_2H_6$ to 3.5 $O_2$ |
| propane | 1 $C_3H_8$ to 5 $O_2$ |
| butane | 1 $C_4H_{10}$ to 6.5 $O_2$ |
| pentane | 1 $C_5H_{12}$ to 8 $O_2$ |

**TABLE 2**: Ratio of alkane molecules to oxygen molecules for combustion reactions.

## QUESTIONS

**6** Describe the pattern of alkane:oxygen ratios for the combustion of alkanes.

**7** Write a balanced equation for the combustion of hexane, $C_6H_{14}$.

Q ... gcse combustion

# Energy resources

## Essential notes

> energy resources may be non-renewable or renewable

> electricity can be generated using fossil fuels (combustion), nuclear fuels (nuclear fission) and renewable energy resources

**THE SCIENCE IN CONTEXT** As you will have heard many times, fossil fuels will not last forever. Even if they did, environmental scientists and others are concerned about the use of fossil fuels because of global warming and pollution issues. The search is on for clean, safe renewable energy sources. Biologists, chemists and physicists are contributing, in different ways, to tackling one of the greatest challenges of the 21st century.

## Solar power stations

Europe's first commercial solar power station began generating electricity in 2007. Hundreds of computer-controlled mirrors move to reflect sunlight onto a tower to heat water. The steam produced drives a turbine.

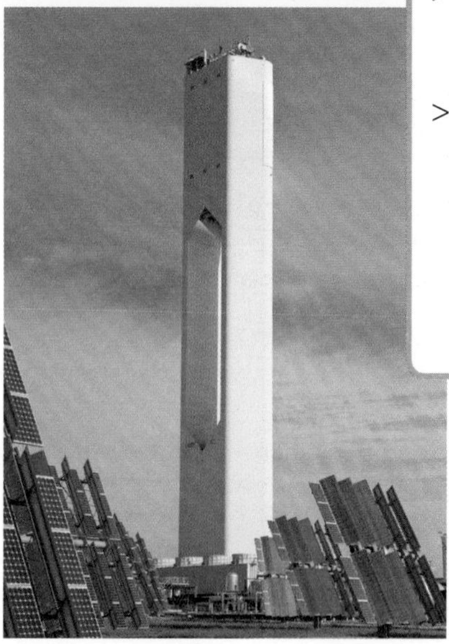

**FIGURE 1**: By 2013, the Seville solar power station in Spain will supply electricity for 180 000 homes.

## Renewable and non-renewable

### Non-renewable energy resources

Fossil fuels and **nuclear fuels** are **non-renewable** energy resources. Once used, they cannot be replaced. Both release energy. Burning fossil fuels releases the energy stored in, mainly, hydrocarbons.

Nuclear fuels are also non-renewable. **Nuclear fission** releases the stored energy in atoms. This happens when the nuclei of atoms are split up. Radioactive isotopes of uranium and plutonium are the usual fuels.

Unlike fossil fuels, nuclear fuels do not produce gases such as carbon dioxide that cause global warming. Nor do they produce polluting gases such as sulfur dioxide. However, waste materials from nuclear fuel power stations are **radioactive** and harmful to life. Waste must be stored safely until its radiation levels are no longer harmful.

Fossil fuels and nuclear fuels are used in power stations to generate electricity.

**FIGURE 2**: Generating electricity from moving water.

### Renewable energy resources

Renewable resources can be replaced – they do not 'run out' – and may be used as alternatives to fossil fuels and nuclear fuels. They include:

> wind

> solar

> hydroelectric

> wave

> tidal

> biomass

> geothermal.

All have their advantages and disadvantages. Environmental impact, reliability and costs are important considerations.

### Did you know?

In 2008, renewable resources supplied 19% of global energy consumption.

## QUESTIONS

1 Name two non-renewable energy resources.

2 Name three renewable energy resources that use moving water.

3 Explain the difference between non-renewable and renewable resources.

## Renewable energy sources

### Wind and water

The energy of wind, rivers (hydroelectric) and the sea (waves and tides) can be harnessed. In each case, the moving air or water is used to turn **turbines**, which turn generators, producing electricity.

None produce carbon dioxide that might contribute to global warming. All are non-polluting.

However, they may be unsightly, or even ugly, in some people's eyes. Their construction and operation may harm wildlife or destroy natural habitats. Wind and waves are not reliable. It may be windy or still. The sea might be rough or calm.

### Biomass

Domestic, industrial or agricultural waste may be burned in 'energy-from-waste' power stations. Much of this is biomass, such as waste woodchips, used animal bedding, nutshells or meat production waste.

Other power stations burn methane gas from landfill sites or sewage, or biofuels. Biofuels specially made from materials such plant oils or fermented sugar cane.

### Solar

In solar power systems, energy from the Sun is harnessed directly. At its simplest, water passes through pipes on the roof of a house, for example, and the Sun heats it directly.

However, to produce electricity you need photovoltaic cells (PV cells). You may have seen panels made of hundreds of these cells on rooftops. The problem is that the Sun does not shine all the time.

### Geothermal

Rocks deep underground are hot. In volcanic areas, steam sometimes rises to the surface. Geothermal power stations use the steam to drive a turbine. In other places, water is pumped through boreholes into the hot rock. Steam comes back and the steam drives a turbine.

FIGURE 3: What is the difference between a 'wind turbine' and a 'wind farm'?

FIGURE 4: Solar panels power this car.

### QUESTIONS

**4** Why is energy from wind and waves unreliable?

**5** Which gas is produced in landfill sites or sewage?

**6** What are the two ways that energy from the Sun can be harnessed?

## Storing energy

The electricity generated from wind or solar energy is much less if there is no wind or the Sun is not shining.

In 2010, engineers from Cambridge suggested giant 'gravel batteries' to store energy. When more electricity is generated than is needed, it is used to heat and pressurise gas. The gas is then pumped though gravel.

The hot gravel stores energy. When needed, cooler gas is passed through the gravel, heats up and can be used to drive a turbine.

### Remember

Fossil fuels release energy by combustion. Nuclear fuels release energy by nuclear fission.

### QUESTIONS

**7** Suggest some reasons why gravel is a good material to use in these 'giant batteries'.

# Electricity generation

THE SCIENCE IN CONTEXT Somebody once said that there were many advances in the 20th century, and most of them plug into the wall. It was a nice way to summarise our growing dependence on electricity. In the 21st century, it is the same. Different methods of generating electricity have differing advantages and disadvantages. Advances in the technology used will be vital to the future of electricity generation.

## Exploding lake

Highly flammable gases from volcanic activity, and from bacteria, are building up at the bottom of Lake Kivu in Rwanda. Scientists fear there may be a massive explosion, destroying local villages.

FIGURE 1: Lake Kivu – In 2003, engineers started pumping out the gas and burning it in a power station to generate electricity.

## Generating electricity

The key to generating electricity in **power stations** is steam. It is the first, vital stage of the process.

The stages in electricity generation are:

1. heat water to produce steam

2. use steam to turn a turbine, returning condensed steam (now water) to the boiler

3. use the spinning turbine to make the generator turn

4. generator generates electricity

5. step-up transformers increase the **voltage** to the very high voltages needed for the **National Grid**.

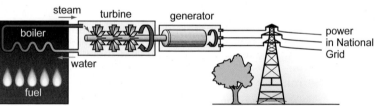

FIGURE 2: Generating electricity.

Steam is produced either by burning fossil fuels or from the energy released in nuclear fission. In 2008:

> fossil fuels were used to generate 69% of global electricity

> nuclear fuels were used to generate 13% of global electricity.

### Transformers

Electricity produced by power station generators is not at a high enough voltage to carry electricity efficiently through National Grid cables (power lines) that distribute electricity generated in power stations to homes and offices.

> **Step-up transformers** are used to produce the very high voltages needed.

However, these high voltages are too dangerous to use in homes.

> **Step-down transformers** are used to reduce the voltage to about 230 V.

### Did you know?

A large power station may burn a trainload of coal every half-hour.

### QUESTIONS

1 Describe what happens in the boiler of a power station.

2 Which part of the power station generates electricity?

3 Explain what a step-up transformer does.

# Efficiency

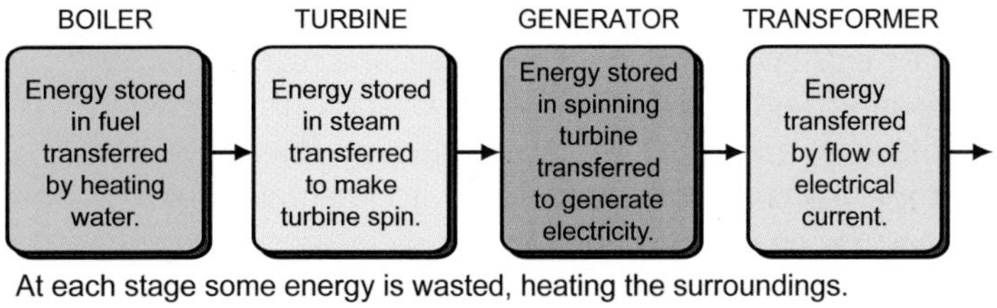

Power stations take energy stored in fossil fuels or nuclear fuels and transfer it into electricity. Energy released from the fuels is the **total energy input**. The law of conservation of energy says that energy can be neither created nor destroyed. So, the **output energy** must equal the input energy. However, it can spread out.

| BOILER | TURBINE | GENERATOR | TRANSFORMER |
|---|---|---|---|
| Energy stored in fuel transferred by heating water. | Energy stored in steam transferred to make turbine spin. | Energy stored in spinning turbine transferred to generate electricity. | Energy transferred by flow of electrical current. |

At each stage some energy is wasted, heating the surroundings.

**FIGURE 3**: Energy transfers in a power station.

> Some energy does what is needed. It produces electricity. This is **useful energy**.

> Some energy heats up machinery and the surroundings. It does not produce electricity. This is **wasted energy**.

The equations are:

input energy = output energy (useful energy + wasted energy)

$$\text{efficiency} = \frac{\text{useful energy output}}{\text{total energy input}} \times 100\%$$

## Power station efficiencies

In a power station, energy is wasted at every stage. The efficiency of traditional fossil fuel power stations is about 35%. Nuclear power stations are about 30% efficient.

Most new fossil fuel power stations being built are 'combined cycle gas turbine power stations'. The first stage uses burning natural gas to drive round one turbine, connected to a generator. Waste energy from this 'gas turbine' is used to heat water. The steam drives a steam turbine. This type of power station is just over 50% efficient.

The most efficient fossil fuel power stations are combined heat and power stations. The waste energy is used to heat local houses and businesses. These are usually 70% to 80% efficient.

In some modern gas-fired power stations, the burning gas is used to heat air instead of water. The hot air is pressurised and used to drive round a turbine.

## QUESTIONS

**4** In a power station, what happens with (a) useful energy (b) wasted energy?

**5** How can the efficiency of a power station be calculated?

**6** Explain why a combined cycle gas turbine power station is more efficient.

# What should we burn?

Large amounts of coal are present in deep underground seams that are inaccessible to mining. In 2002, some engineers suggested that it would be possible to burn this coal. The resulting gases, trapped at the surface, could fuel gas turbine power stations.

Critics say the resulting fires in the coal seams would probably be impossible to put out. Also, the burning coal would release massive amounts of carbon, possibly causing unstoppable global warming, affecting the whole **planet**.

## QUESTIONS

**7** Discuss the plan to burn coal in deep underground seams. What extra information do you need?

**FIGURE 4**: Which fuels should be burned?

# Electricity distribution

**THE SCIENCE IN CONTEXT** Generating electricity is one challenge. Getting it to homes and places of work is another challenge. The National Grid is the network of cables and wires that makes this possible. However, there are concerns about the environmental problems and possible health risks of distributing electricity over the land by pylons and high-voltage cables.

## Electricity without wires

Imagine recharging your mobile without plugging it into the wall. The technology is there. It needs further development, but electricity without wires might be part of your future.

**FIGURE 1**: Are there invisible risks here?

## The National Grid

Mains electricity is generated in power stations. The network of cables that carries electricity to local areas is called the National Grid. Electricity companies then run smaller cables to your school and home.

Electricity from power stations carries very large amounts of energy. This is because one power station may supply the energy for thousands of homes and businesses. It makes the cables extremely dangerous. You may have seen warning signs saying 'Danger Overhead Cables'. Near canals or lakes the sign is to warn against accidentally touching cables with fishing rods – an electric current would flow down the rod and kill the fisherman.

### Changing voltage

Electricity from a power station has a very high voltage. Most of it is distributed around the country at 400 000 V, but the mains supply in our homes is only at 230 V. Step-down transformers, in large sub-stations, reduce the voltage to 132 000 V. Then more step-down transformers, in smaller sub-stations, reduce it to 230 V. There is usually a small sub-station for every few streets.

**FIGURE 2**: Why it is dangerous to fly a kite near overhead cables?

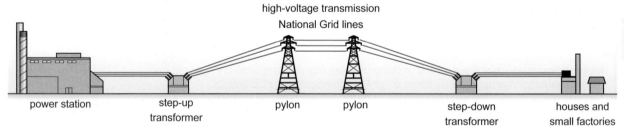

high-voltage transmission
National Grid lines

power station    step-up transformer    pylon    pylon    step-down transformer    houses and small factories

**FIGURE 3**: The National Grid distributes electricity throughout the UK.

## QUESTIONS

1 What is the National Grid?

2 Explain why electricity, as it leaves a power station, has high energy.

3 Describe what happens at an electricity sub-station.

### Did you know?

The National Grid consists of 22 000 pylons, 4500 miles of overland cable and 420 miles of underground cable.

# High voltages?

**Electrical power** is the energy transferred by an electric current each second.

This **power** depends on the size of the current and the voltage.

power = current × voltage, or $P = I \times V$

The National Grid distributes large amounts of power, and so $P$ in the formula is high. The current can be kept low as long as the voltage is high. If the current is small, the wires carrying it can be thinner. Thinner wires contain less metal, so are cheaper.

A huge current does not come into homes when the voltage is changed down to 230 V. The single current from the power station splits into lots of smaller, parallel currents, all taking electricity to different houses.

## Heat losses

Some of the energy carried by an electric current is wasted. Wires become warm. As the current increases, the amount of energy that heats the wires increases. Keeping the current from the power station low, wasted energy is reduced. About 10% of the energy from power stations is wasted in the National Grid.

> **Remember**
> Very high voltages transmit very large amounts of energy with relatively small energy losses.

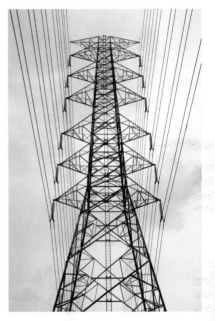

## QUESTIONS

**4** If the voltage from a power station doubled, how would the current change?

**5** Give one advantage of using thin wires to transfer energy, other than cost.

**6** If 10% of the energy is wasted, what is the efficiency of the National Grid?

FIGURE 4: What advantages are there in using high voltages in these cables?

# Choosing cables

Most of the connecting wires you have used in circuits are copper, with plastic insulation around them. Copper is a better electrical conductor than most metals, so a smaller proportion of the energy flowing through it is wasted. In the National Grid, overhead transmission cables are aluminium reinforced with steel, not copper. Also, the cables are not insulated.

## QUESTIONS

**7** Explain why National Grid cables, on pylons, are not insulated and whether or not this affects safety.

**8** Suggest some factors that affect the choice of putting cables overhead or underground.

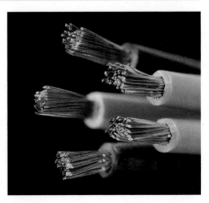

FIGURE 5: A cable is a bundle of wires.

# Unit 2 Theme 2 Checklist

## To achieve your forecast grade in the exam you will need to revise

Use this checklist to see what you can do now. Refer back to the relevant pages in this book if you are not sure. Look across the three columns to see how you can progress. **Bold** text means Higher tier only.

Remember that you will need to be able to use these ideas in various ways, such as:

> interpreting pictures, diagrams and graphs

> applying ideas to new situations

> explaining ethical implications

> suggesting some benefits and risks to society

> drawing conclusions from evidence that you are given.

Look at pages 250–271 for more information about exams and how you will be assessed.

| To aim for a grade E | To aim for a grade C | To aim for a grade A |
|---|---|---|
| Recall that limestone is obtained from the ground by quarrying. Give uses of limestone in the building industry. | Describe the composition and use of mortar and concrete. Outline the manufacturing processes for quicklime, cement and glass. | Describe the conversion of limestone into quicklime and quicklime into slaked lime. Know the chemical formulae for quicklime and slaked lime. |
| Recall the characteristic properties of metals. | Relate uses of metals in the building industry to the properties of these metals. | |
| Know that most polymers are manufactured using chemicals obtained from crude oil. | Describe how polymers are produced when monomers join together to form very large molecules. | Know the properties of polymers that relate to their uses in the home. |
| Describe the properties of a composite as a combination of the properties of its components. | Relate the properties of ceramics to their uses in construction. Recognise and describe a composite material. | |
| Recall that hydrocarbons contain carbon and hydrogen only. Know that the energy in hydrocarbons is released when they are burned in air, which makes them useful as fuels. | Name suitable fuels for cooking, heating our homes and for providing transport. Know that crude oil is an important source of fuels used for cooking and heating and for transport. | Discuss why environmental scientists are concerned about the use of fuels obtained from crude oil. |

| To aim for a grade E | To aim for a grade C | To aim for a grade A |
|---|---|---|
| Know that resources of fossil fuels are finite. | Explain some of the problems of burning fossil fuels.<br><br>Write word equations for the combustion of hydrocarbons.<br><br>Discuss the impacts of the uses of fuels obtained from crude oil. | **Write balanced symbol equations for the complete combustion of hydrocarbons.**<br><br>**Recognise the pattern in chemical formulae based on $C_nH_{2n+2}$.**<br><br>**Recognise qualitative and quantitative patterns in the amounts of reactants and products.** |
| Know that fossil fuels release energy when they are burned, which can be used to generate electricity. | Describe how electricity can be generated from fossil and nuclear fuels. | Discuss the advantages and disadvantages of different methods of generating electricity. |
| Define the terms renewable and non-renewable in the context of energy sources. | Know that nuclear fuels produce energy from nuclear fission.<br><br>Explain how nuclear fuels and renewable energy sources may be used as alternatives to fossil fuels. | Explain the problems of using nuclear fuels and of using renewable energy sources. |
| Describe how electricity is distributed through the National Grid via high-voltage cables. | Describe the use of step-up and step-down transformers. | Explain why people are becoming more concerned about the environmental problems and possible health risks of distributing electricity using pylons and high-voltage cables. |

# Unit 2 Theme 3: My property

## What you should know

### Electricity

Energy can be used in a wide variety of ways.

When energy is used it does not disappear.

Electricity can be generated using a variety of energy sources.

- List some household appliances that are powered by electricity.

### Electromagnetic waves

A prism may be used to disperse white light to give a spectrum of colours.

Frequency is the number of waves transferred per second and is measured in hertz (Hz).

- What does wavelength mean?

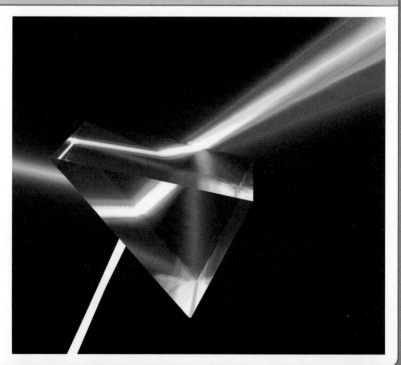

# You will find out

## The cost of running appliances in the home

> Energy is normally measured in joules.

> Power is the rate of energy transfer; one watt is one joule per second.

> Electricity meters measure energy transferred in kilowatt-hours.

> The total cost of running an appliance
= number of kilowatt-hours × cost per kilowatt-hour.

> Sankey diagrams can be used to show energy transfers.

> The efficiency of an electrical appliance
= useful energy transferred ÷ total energy supplied.

## Electromagnetic waves in the home

> Electromagnetic radiation travels as waves, which move energy from one place to another.

> The electromagnetic spectrum ranges from radio waves to gamma rays.

> Waves of shorter the wavelength or higher frequency transfer more energy.

> Velocity (m/s) = frequency (Hz) × wavelength (m).

> Radio waves, microwaves, infrared, visible light and ultraviolet are used in our homes.

> X-rays and gamma rays are used in medicine.

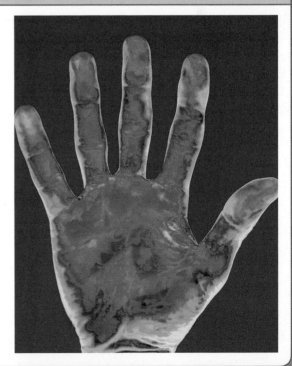

# Electrical appliances

### THE SCIENCE IN CONTEXT

Electrical appliances, devices, gadgets and machines are part of our everyday life. They are everywhere, running off mains electricity or powered by ever smaller, more powerful batteries. Electricians and others who work with electrical appliances need a sound grasp of the ideas of voltage, current and power. They also need to be able to calculate the electricity used in a home or business.

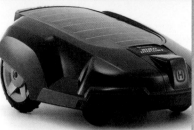

**FIGURE 1**: This mower has a power rating of 32 W.

**Essential notes**

> energy is normally measured in joules
> power is the rate of energy transfer (power = potential difference × current)
> 1 watt = 1 joule per second
> energy transferred = power × time; electricity meters use the unit kilowatt-hours to measure energy transferred

**Green lawns**

This robot lawnmower uses energy from sunshine to provide its power. Photovoltaic panels collect energy radiated by the Sun and generate electricity.

## Voltage, current and power

Electrical appliances are all around you. Close your eyes and list them in your mind. It is not easy to imagine life without them. Some are powered by batteries, others by mains electricity. How much does it cost to run them? To answer this question, you need to know how they transfer energy.

Energy is normally measured in joules (J).

Power is the rate at which an **appliance** transfers energy. It is measured in **watts** (W) – one watt transfers one **joule** of energy each second (J/s).

The power an appliance consumes depends on potential difference and the current it draws from the electricity source (for example, mains electricity or a battery). The equation that you can use is:

power (watts) = potential difference (volts) × current (amps)

In the UK, mains electricity is 230 volts. Batteries provide much lower voltages.

> A calculator uses a 6 volt battery. When in use, a current of 0.08 amps passes through it.

The power it consumes is 6 × 0.08 = 0.48 watts.

> An electrical fan heater uses 230 volts and draws a current of 8.7 amps

The power it consumes is = 230 × 8.7 = 2000 watts (2 kW).

**Remember**
1kW = 1000 W

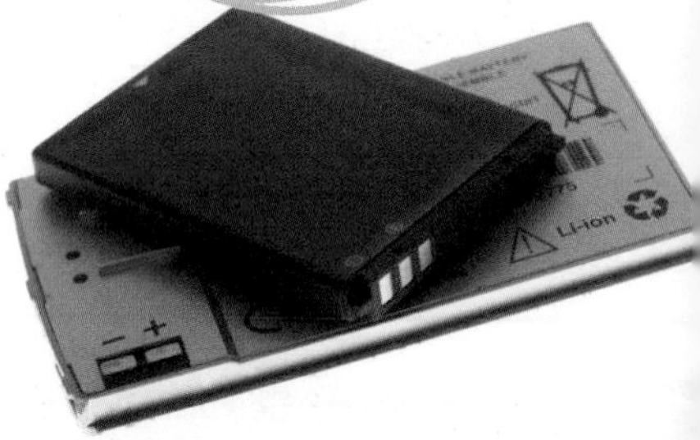

**FIGURE 2**: Rechargeable batteries power mobile phones and many other electrical devices.

**Did you know?**

Current is measured in amps, which is short for ampere. The symbol is A.

| Appliance | Power consumption |
|---|---|
| laptop computer | 100 W |
| electric kettle | 2.4 kW |
| hairdryer | 1.5 kW |
| television | 50 W |
| microwave oven | 1.2 kW |
| electric fire | 2.0 kW |

**TABLE 1**: Power consumption of some household appliances.

### QUESTIONS

1 Explain what the power consumption of an appliance means.

2 Which appliance listed in Table 1, transfers energy at the greatest rate?

3 Calculate the power consumption of an electric drill that draws a current of 1.25 amps from mains electricity.



... gcse current voltage power

## Power calculations

The electricity used in a home or business is measured by an electricity meter. It measures the energy transferred to the various electrical appliances that are used.

The equation that connects power, energy transferred and time is:

> power = energy transferred ÷ time

Two sets of units may be used (see Table 2). Because of the large amounts of power used in homes, domestic electricity meters measure in kilowatt-hours (kW).

| Power | Energy transferred | Time |
|---|---|---|
| watt, W | joule, J | second, s |
| kilowatts, kW | kilowatt-hour, kWh | hour, h |

**TABLE 2**: Two sets of units. Use one set or the other and the correct units.

Suppose an electric fire transfers 12 kWh of energy in five hours. You can calculate the power consumed:

> power = energy transferred ÷ time = 12 ÷ 5 = 2.4 kW

This is the **power rating** of the fire.

**FIGURE 3**: The meter shows that 39 513 kWh of electricity were used since it was installed.

**Remember**
The power an appliance consumes is often called its power rating.

**Remember**
One unit of electricity is one kilowatt-hour (1 kWh).

### QUESTIONS

**4** What does a domestic electricity meter measure?

**5** Calculate the power consumption of a television that transfers 0.9 kWh of energy in 10 hours.

**6** Calculate the power consumption of a hairdryer that transfers 0.05 kWh of energy in three minutes.

## Rearranging the power formula

The equation

power = energy transferred ÷ time

can be rearranged as:

energy transferred = power × time

This allows you to calculate the energy transferred by an appliance.

Suppose a 1.2 kW microwave oven cooks food for three minutes. To find out the total energy transferred in kilowatt-hours:

1. change the values into the correct units

   1.2 kW

   3 minutes = 3 ÷ 60 = 0.05 hours

2. use the equation energy transferred = power × time

   energy transferred by the microwave oven = 1.2 × 0.05 = 0.06 kWh

### QUESTIONS

**7** Calculate the energy transferred by a laptop computer (power rating 100 W) when it is used for 10 hours.

Q ... gcse calculating power

# Electricity costs

**THE SCIENCE IN CONTEXT** Electrical appliances transfer energy for heating and to power electrical appliances in the home. People buying electrical appliances may want to compare not only the prices of a range of similar appliances, but also their running costs. They will also want to know about safety and other features.

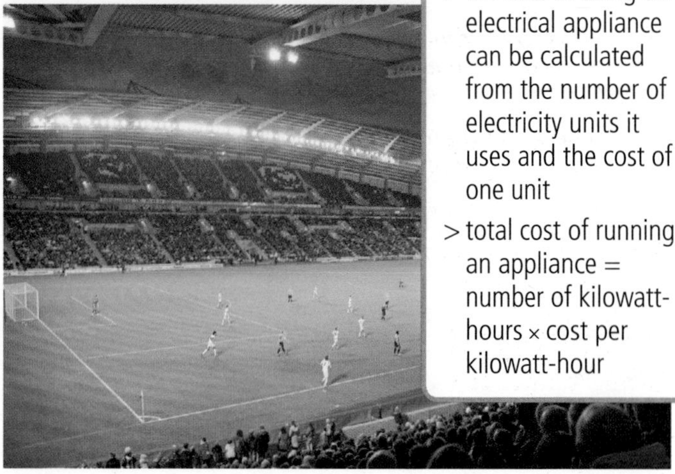

FIGURE 1: Try to estimate the electricity bill for a 90-minute match.

### Essential notes

- the cost of using an electrical appliance can be calculated from the number of electricity units it uses and the cost of one unit
- total cost of running an appliance = number of kilowatt-hours × cost per kilowatt-hour

### Sport under floodlights

A floodlit football match needs about five million watts of electricity – the same amount as a town of 5000 people. It's a huge electricity bill!

## Electricity bills

Electricity costs money. A domestic electricity meter measures how much is used in a home and people are sent an electricity bill for that amount of electricity.

Electricity companies charge for the number of kilowatt-hours of energy transferred.

One unit of electricity = 1 **kilowatt-hour** (kWh)

Electricity meters measure the number of electricity 'units' used.

You can find the cost of using an appliance by using:

cost = number of kilowatt-hours × cost per kilowatt-hour

Suppose you heat a ready-meal in a 1.2 kW microwave oven for 5 minutes. Using:

energy transferred = power rating × time

you can calculate the energy transferred:

$$1.2 \times \frac{5}{60} = 0.1 \text{ kWh}$$

This is 0.1 units on an electricity bill.

If each unit (kilowatt-hour) of energy costs 12p, the cost of heating the ready-meal will be:

0.1 × 12 = 1.2p

To calculate the cost of using several appliances, you find the total energy used by them (total kilowatt-hours) and multiply this by the cost of a unit of electricity.

total cost = total kilowatt-hours × cost per kilowatt-hour

**LANCASHIRE** *ELECTRIC*

Mr Pritchard
74 Green Avenue
Preston
LANCASHIRE

***Account number*** 6645/3526/936B

***Date*** 30 June 2010

***Electricity bill***

| | £ | £ |
|---|---|---|
| Previous bill 30 March 2010 | 224.67 | |
| Payment received 7 April 2010 | 224.67 CR | |
| Balance brought forward | | 0.00 |
| Present electricity charges | | 197.04 |
| VAT @ 5% on present charges | | 9.85 |
| | TOTAL | 206.89 |

***Payment Due*** by 25 July 2010 **£206.89**

***How your bill is calculated***

| Meter number | METER READINGS Present | Previous | Units used | Rate | Unit cost | Charge £ |
|---|---|---|---|---|---|---|
| 06039 | 45698 | 44096 | 1602 | Normal | 11.52p | 184.55 |
| | Standing charge 30 Mar 10 – 30 Jun 10 | | | | | 12.49 |
| | | | | | Present electricity charges | 197.04 |

FIGURE 2: On this electricity bill, how is the number of 'Units used' calculated?

### QUESTIONS

1 If electricity costs 12p per kilowatt-hour, how much would it cost to cut a large lawn that takes 2 hours to mow, with a 1.0 kW mower?

2 To mix some concrete, one mixer, rated at 600 W, takes 1 hour. A different mixer, rated at 400 W, takes 2 hours. Explain which is cheapest to use.

## Kilowatt-hours

Electricity supply companies use kilowatt-hours because it is more convenient than joules – an average electricity bill would be millions of joules.

For example, using a 2 kilowatt oven for an hour means that more than seven million joules of energy is transferred. Here is the calculation:

2 kilowatts = 2000 watts (2 kW = 2000 W)

1 hour = 60 × 60 = 360 seconds (1 h = 360 s)

Using:

energy transferred (J) = power (W) × time (s)

energy transferred = 2000 × 360 = 7 200 000 J

**FIGURE 3**: Listening to music uses energy. Who pays?

The kilowatt-hour is the same type of unit as a **joule**, but it is used to for larger quantities of energy. The formula for calculating the energy is the same, but with different units:

energy transferred (kWh) = power rating (kW) × time (h)

Using a 2 kilowatt oven for an hour:

energy transferred = 2 × 1 = 2 kWh

So, 2 kWh and 7 200 000 J are the same amounts of transferred energy, but 2 is a more manageable number than 7 200 000.

**Remember**

To work out the cost, power must be in kilowatts and time in hours.

### QUESTIONS

**3** If a 1 kW electric fire is left on for 1 hour, how much energy does it use in (a) kWh (b) J?

**4** Calculate the kWh that 10 laptops, rated 100 W each, use in one hour.

**5** A plasma screen television has a power rating of 200 W. How much energy does it use, in kWh, if left on for two hours?

**Remember**

A watt is the unit of rate of energy transfer. 1 watt = 1 joule per second (1 W = 1 J/s)

## Economy 7

Using electricity at night can be cheaper than during the day. Economy 7 operates typically from midnight and lasts for 7 hours. It is roughly one-third the cost of daytime electricity. Generally speaking, you need to use more than 40% of your electricity at night to make Economy 7 cost effective.

### QUESTIONS

**6** Find out (a) why Economy 7 is cheaper than daytime electricity (b) the tariff for Economy 7 where you live. Try to decide if Economy 7 would be worth having in your home.

**Did you know?**

Playing a computer game for an hour uses the same amount of energy as boiling the water for one cup of coffee.

# Efficiency of appliances

**THE SCIENCE IN CONTEXT** People buying electrical appliances may want to compare the efficiency of appliances, for example, washing machines or refrigerators. Electrical appliances transfer energy, but some do it more efficiently than others and would be more cost-efficient to run. Nowadays, many electrical appliances have energy labels to help consumers decide which are most efficient and cost-effective.

## Saving energy

Older light bulbs have a tungsten filament. They use only 10% of the energy put into them to give light. The rest is 'wasted' by heating the surroundings. Filament bulbs are being replaced by 'low energy' light bulbs. In contrast to filament bulbs, low energy light bulbs use 40-50% of the energy to give light.

**FIGURE 1**: A tungsten filament bulb and a 'low energy' bulb.

## Energy transfers

Mains electricity is convenient, clean, safe and reliable. You just flick a switch and there is the energy when you need it. A wide range of appliances transfer the energy supplied by electricity into something useful. For example, a light bulb provides light and an MP3 player provides sound. The energy transferred to do these things is **useful energy**.

In all appliances, some energy does not do useful work. The purpose of an electric light bulb is to produce light. However, it also gives out heat when being used. Energy that does not produce light is **wasted energy**.

*Energy can be neither created nor destroyed.* This is the **Conservation of Energy** law.

Whatever energy is put into an appliance, the same amount must come out. This can be written as:

energy input =
useful energy output + wasted energy output

The greater the proportion of useful energy output, the more efficient the appliance.

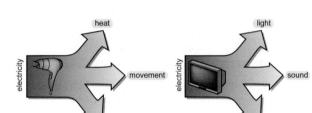

= useful energy transfers     = wasted energy transfers

**FIGURE 2**: Energy transfer in a hairdryer and in a television.

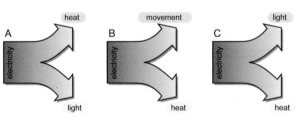

= useful energy transfers     = wasted energy transfers

**FIGURE 3**: Match the diagrams to these appliances: light bulb, electric fire, motor.

### Did you know?

EU Energy labels indicate the efficiency of electrical appliances, such as refrigerators, cookers and washing machines.

## QUESTIONS

**1** Explain the difference between useful energy and wasted energy.

**2** Why must the useful energy output and wasted energy output equal the input energy to an appliance?

**3** Describe what is meant by an 'efficient appliance'.

Q ... energy transfer in electrical appliances

# Sankey diagrams

A **Sankey diagram** shows energy transfers that happen in a device. It shows the energy:

> supplied to an appliance (energy input)

> transferred from an appliance to do different things (energy output).

The widths of the arrows are drawn to scale. They show the relative proportions of input and output energy. The values are given on the diagram.

The total amount of energy stays the same (the law of Conservation of Energy).

## Efficiency

An efficient person is someone who gets the work done without lots of fuss and without using energy unnecessarily. An efficient electrical appliance is similar – it transfers as much energy as possible to do what it is designed for.

You can calculate efficiency using:

$$\text{efficiency} = \frac{\text{useful energy transferred by the appliance}}{\text{total energy supplied to the appliance}}$$

Figure 4 shows Sankey diagrams for a filament light bulb and a low energy light bulb. You can calculate the efficiency of each using the equation above.

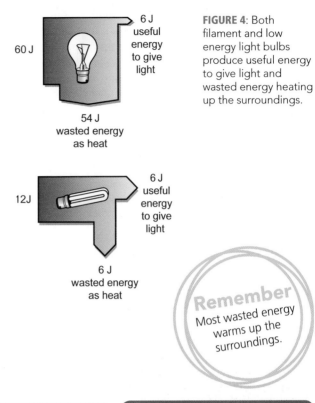

FIGURE 4: Both filament and low energy light bulbs produce useful energy to give light and wasted energy heating up the surroundings.

**Remember**
Most wasted energy warms up the surroundings.

|  | Filament bulb | Low energy bulb |
|---|---|---|
| total energy input | 60 J | 12 J |
| useful energy output to give light | 6 J | 6 J |
| efficiency | $\frac{6}{60} = 0.1$ | $\frac{6}{12} = 0.5$ |
|  | *To convert to a percentage, multiply by 100...* | |
|  | 10% | 50% |

Efficiency may be given as a decimal fraction or a percentage.

## QUESTIONS

**4** If the total input energy of an appliance is 8000 J and the wasted energy output is 2000 J, what is the efficiency of the appliance?

**5** Explain why an efficient appliance costs less to run than an inefficient appliance.

# Energy labels

Energy labels tell you about the efficiency and energy consumption of electrical appliances from energy labels. Efficiency is graded from A to G, but it is not calculated using the equation that you used earlier. How it is calculated varies between types of appliances. Recently A++ and A+ grades have been introduced for some types of appliances.

## QUESTIONS

**6** Look at the EU energy label in Figure 5. In a local store, find out how this machine compares with others available. How can energy labels help you decide which one to buy?

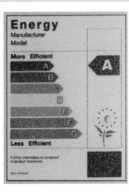

FIGURE 5: An EU energy label for a washing machine.

# Electromagnetic radiation

**THE SCIENCE IN CONTEXT** Electronic engineers use electromagnetic radiation for radio, mobile phones, and cable and satellite television. Electromagnetic radiation transfers energy from one place to another. The energy is carried by electromagnetic waves, which can move through a vacuum. This is why satellite communications can work.

### Internet access in remote places

Even without phone lines, cables or wireless hotspots, you can still access the internet using a satellite connection. Signals are sent as electromagnetic waves, which travel though space.

FIGURE 1: You can go online using a smartphone.

## Waves

Waves move energy from one place to another. There are different types of waves, but they fall into two groups:

> **mechanical waves** – waves in the sea are mechanical

> **electromagnetic waves** – waves that transfer some of the Sun's energy to Earth are electromagnetic.

**Electromagnetic radiation** moves energy by **electromagnetic waves**. Cable and satellite TV, mobile phones and radio signals are all sent using electromagnetic waves. You cannot see electromagnetic waves – only their effects.

It is helpful to have a 'picture' in your mind of waves. One way is by thinking about waves in a piece of rope. Tie the end of a piece of rope to a table leg. Hold the other end and move the rope up and down. A wave moves along the rope.

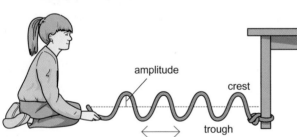

FIGURE 2: Making a wave by waggling a piece of rope.

The rope wave is an example of a mechanical wave. Waves in the sea or ripples on a pond are other examples. All need a material to transfer the energy (through the rope or water).

Electromagnetic waves are different. They do not need a material to transfer the energy. It is the electrical or magnetic fields that vibrate as the waves travel through them – so electromagnetic waves can pass through a **vacuum**.

### Did you know?

The most beautiful red sunsets are caused by tiny dust particles in the atmosphere, often from erupting volcanoes.

### QUESTIONS

**1** Describe the meaning of (a) wavelength (b) amplitude.

**2** Describe something that all waves have in common.

**3** Sound cannot travel through a vacuum. Why not?

Q ... gcse electromagnetic waves

# Wavelength

The wave in Figure 2 has a characteristics called wavelength. It is the distance between one crest and the next (or one trough and the next).

If you waggle the rope more vigorously, you will see more waves. The distance between crests is shorter. The waves have a shorter wavelength. The more energy you transfer to the rope, the shorter the wavelength of the waves you make. This is a very important idea.

Now, imagine a Mexican wave (you may even have taken part in one at some time). For this human wave:

> a crest is where the person is standing fully upright with their arms in the air

> a trough is where the person is seated.

If the wave is done lazily and slowly, there might be 50 people between each crest.

If the wave is done energetically and quickly there might be only 20 people between each crest.

The more energy put in, the shorter the wavelength of the human wave will be.

It is the same with electromagnetic waves.

The more energy put into their formation, the shorter their wavelength will be.

Infrared radiation has waves of longer wavelength than waves of **ultraviolet radiation**. They carry less energy. This is why infrared radiation does not cause sunburn, but ultraviolet radiation can.

FIGURE 3: You could estimate the wavelength of a human wave by counting the number of people between two that are standing fully upright with their arms in the air.

## QUESTIONS

**4** How could you measure the wavelength of a Mexican wave?

**5** Describe the relationship between wavelength and the energy put in to create a wave.

FIGURE 4: Ultraviolet radiation can tan you, but it can also damage your skin.

# The visible spectrum

White light is electromagnetic radiation. It is carried by waves of different wavelengths. These can be separated by passing the light through a glass prism. The waves are diffracted (change direction) when they travel from air into to glass and then again from glass into air (Figure 5).

## QUESTIONS

**6** Describe the relationship between wavelength and the amount that a wave is diffracted by a glass prism.

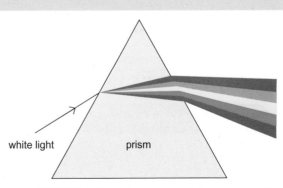

white light    prism

FIGURE 5: White light consists of electromagnetic waves that have a range of wavelengths.

# Wavelength, frequency, energy

**THE SCIENCE IN CONTEXT** The electromagnetic spectrum ranges from radiowaves to gamma rays. Radiation in all parts of this spectrum is used for a range of applications. The amount of energy transferred by electromagnetic waves depends on their wavelength and frequency. To make use of electromagnetic radiation, scientists need to know how it affects materials, including the human body.

## SmartWater

SmartWater can be painted on any surface. When dry, it is almost invisible to the naked eye. However, it glows in ultraviolet radiation. It rubs off on anyone who handles items with it. Shine ultraviolet on their hands and the criminal is linked to the crime.

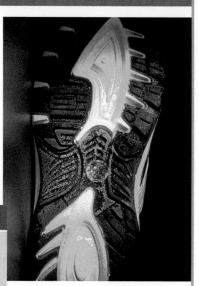

**FIGURE 1**: A burglar was convicted when police officers found traces of SmartWater on his trainers.

## The electromagnetic spectrum

Electromagnetic (EM) radiation travels as waves. It carries energy from place to place. Electromagnetic waves move in straight lines. They travel at the same speed in a vacuum – often called the **speed of light**.

All electromagnetic radiation transfers energy through a vacuum at the same speed. Different types of electromagnetic radiation transfer different amounts of energy. The **electromagnetic spectrum** shows the different types of electromagnetic radiation in the order of the amount of energy that they transfer.

### Did you know?

The wavelength of radio waves is about 10 centimetres (0.1 m). The wavelength of gamma waves is about 10 picometres (0.00000000001 m).

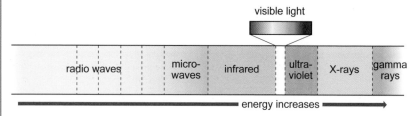

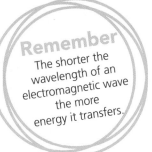

**Remember**
The shorter the wavelength of an electromagnetic wave the more energy it transfers.

**FIGURE 2**: The electromagnetic spectrum. Remembering the order will help you to understand the way different radiations behave.

## QUESTIONS

**1** How does electromagnetic radiation travel, how fast does it go and what does it carry?

**2** Describe what the electromagnetic spectrum shows.

**3** Which type of radiation has the highest energy?

Q ... gcse wavelength ... gcse electromagnetic spectrum

# Frequency

Picture what happens when the participants in a Mexican wave put more energy into it – they stand up and sit down more quickly. The wavelength gets shorter.

You can measure the wavelength by freeze-framing the action and measuring the distance from one person standing to the nearest person who is also standing.

Imagine that instead of measuring the wavelength you look at just one person. Start a stopwatch when she is sitting down. She stands up and puts her arms in the air. Then she sits down again. Stop the watch. The time shown on your stopwatch is how long it has taken for her to go through one complete cycle – from sitting to standing and back to sitting again. You have measured the frequency of the wave.

In scientific language, frequency is the number of waves passing a particular point in a second. It is measured in cycles per second. This unit is called a hertz (Hz).

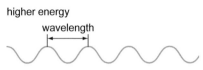

**FIGURE 3**: Wavelength, frequency and energy are linked.

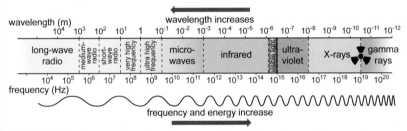

**FIGURE 4**: The electromagnetic spectrum with wavelengths and frequencies.

In the electromagnetic spectrum, waves that transfer:

> more energy have a shorter wavelength and higher frequency

> less energy have longer wavelength and lower frequency.

## QUESTIONS

**4** List the different types of electromagnetic radiation in descending order of frequency.

**5** Explain why electromagnetic radiation of short wavelength has a higher frequency than electromagnetic radiation of longer wavelength.

# Calculations

The frequency and wavelength of any wave are connected by the equation:

velocity (m/s) = frequency (Hz) × wavelength (m)

This may be rearranged as:

$$\text{frequency (Hz)} = \frac{\text{velocity (m/s)}}{\text{wavelength (m)}}$$

or

$$\text{wavelength (m)} = \frac{\text{velocity (m/s)}}{\text{frequency (Hz)}}$$

All electromagnetic radiation travels at the speed of light: $3 \times 10^8$ m/s (300 000 000 m/s).

The wavelength of visible light is approximately $5 \times 10^{-7}$ m, so its frequency is:

$$\frac{3 \times 10^8}{5 \times 10^{-7}} = 0.6 \times 10^{15} = 6 \times 10^{14} \text{ Hz}$$

The frequency of **microwaves** is approximately $1 \times 10^{11}$ Hz, so their wavelength is:

$$\frac{3 \times 10^8}{1 \times 10^{11}} = 3 \times 10^{-3} \text{ m}$$

## QUESTIONS

**6** The wavelength of longwave radio is approximately $1 \times 10^4$ m. Calculate its frequency.

**7** The frequency of gamma rays is approximately $1 \times 10^{20}$ Hz. Calculate their wavelength.

# Using electromagnetic waves

**THE SCIENCE IN CONTEXT** Depending on their wavelength, frequency and wave energy, different types of electromagnetic radiation can be put to many uses. Examples are: mobile phones, television remotes, medical imaging techniques and cancer treatment. Some people have concerns about the risks of using devices that rely on electromagnetic waves.

## Mobiles and health

Some people are concerned that mobile phone use may damage health. In 2010, the chief medical officer for Wales urged schoolchildren to text rather than call on their mobiles. Interestingly, in the same year, scientists in Florida suggested that mobile phone radiation might protect against Alzheimer's disease.

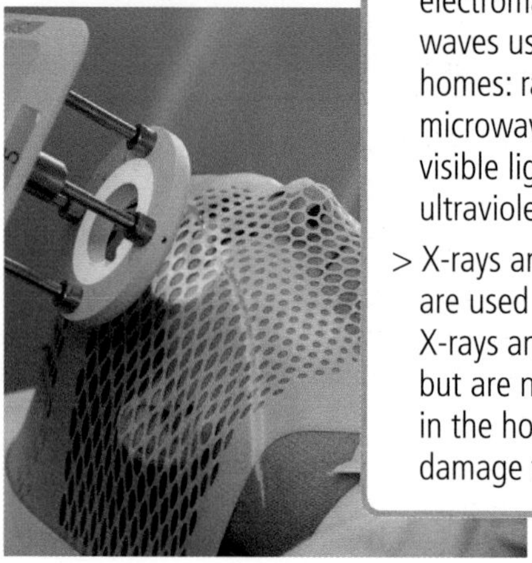

**FIGURE 1**: Gamma rays can used to identify and to treat brain tumours.

### Essential notes

> different types of electromagnetic waves used in our homes: radio waves, microwaves, infrared, visible light and ultraviolet (UV)

> X-rays and gamma rays are used in medicine for X-rays and radiotherapy, but are not usually used in the home as they can damage the body

## Invisible waves

Electromagnetic radiation is invisible. It travels in waves at the speed of light and transfers energy. It has a range – a spectrum – of energies.

Radiation from each part of the electromagnetic spectrum has a number of useful applications.

Increasing energy ↓

**Radio waves** are low energy. They are used to send radio, TV and GPS signals.

**Microwaves** are used in microwave ovens, communication systems and speed traps. Mobile phones and satellite TV use microwave signals bounced off satellites in space.

**Infrared** is used in TV and DVD remote controls. Electric fires emit infrared radiation that warms the room.

**Visible light** enables us to see, including with spectacles, microscopes and telescopes. It is also used in fibre optic cables.

**Ultraviolet** tans skin. Sun-beds use UV light to tan skin artificially.

**X-rays** have high energy. They are used to produce X-ray images of bones and teeth.

**Gamma rays** have very high energy. They can kill cancer cells.

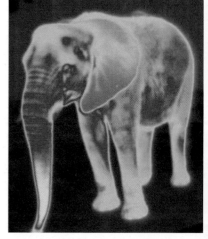

**FIGURE 2**: A thermogram reveals infrared radiation that you cannot see.

Waves within each region of the electromagnetic spectrum have a range of wavelengths, frequencies and energies. The electromagnetic spectrum is continuous. There is not an exact border line between two neighbouring regions.

### Did you know?

Bluetooth signals that enable mobile phones to communicate with laptops or landline phones are very weak radio signals.

### QUESTIONS

1 List the different types of radiation. Give an example of how each is used in a helpful way.

2 What are your 'top three' most useful applications?

3 Explain the meaning of a continuous spectrum.

# Using waves

### Radio waves

Like waves in all parts of the electromagnetic spectrum, radio waves have a range of energies. The lowest energies transmit local radio and police and ambulance messages. The highest energies are for TV.

### Infrared, visible and ultraviolet radiation

Anything warm emits infrared radiation. Infrared cameras detect it and produce thermogram images. For example, thermograms can:

> show the presence of arthritis or cancer in the body, because these parts of the body are warmer

> 'see' a person trapped under rubble or breaking into a house at night.

Visible light is emitted from the Sun, fluorescent lamps and some lasers. Really, it is invisible. It is called visible because you can see where it comes from and the objects that reflect it.

Ultraviolet (UV) radiation is emitted from the Sun and very hot objects such as welding equipment. Sunscreens block some of the Sun's harmful UV radiation that can damage skin.

### X-rays and gamma rays

X-ray images show when bones are broken, teeth decaying or lungs diseased. However, **X-rays** can damage body cells and must be used carefully. This is why radiographers stand behind protective screens when taking X-ray images.

Gamma rays are produced by radioactive substances. These include uranium and its compounds. Gamma rays are used to sterilise food by killing **bacteria**, to kill cancer cells, and in industry to check for faults and cracks in pipelines.

**FIGURE 3**: What types of waves are used to detect speeding motorists?

**FIGURE 4**: UV radiation is useful for checking bank notes for forgeries.

## QUESTIONS

**4** How can infrared rays be used to find people trapped in collapsed buildings?

**5** In a group, discuss safety measures that must be taken when using gamma rays.

**6** What property of gamma rays makes them suitable for sterilising food?

# Security

Airport security ensures millions of passengers fly in safety. You put your baggage on a conveyor belt, where it is X-rayed. Airport staff see an **image** of the object on a screen. Passport control staff check use UV radiation to identify illegal passports.

An infrared burglar alarm senses intruders from their body heat. A laser beam can also be used as a burglar alarm, acting as an invisible light 'trip wire'. If the beam is broken it sets off the alarm.

## QUESTIONS

**7** A valuable diamond necklace is on display at an exhibition. Suggest how you might protect the necklace.

**FIGURE 5**: A gun is detected in a suitcase.

Q ... x-rays ... gamma rays

# Preparing for assessment: Applying your knowledge

*To achieve a good grade in science, you not only have to know and understand scientific ideas, but you need to be able to apply them to other situations and investigations. These tasks will support you in developing these skills.*

## ✳ Getting hot in the kitchen

Consumers have a wide range of appliances to choose from when they cook food or make hot beverages in the kitchen. Most kitchens, for example, have a conventional oven and hob (electric or gas) and a microwave oven.

Defra (Department for Environment, Food and Rural Affairs) produced a briefing note comparing microwave ovens and traditional electric ovens and hobs for cooking. It is part of Defra's *Market Transform Programme* which supports the UK Government policy on sustainable products.

The briefing note contains details of various tests. Defra concluded there was "potential for energy savings to be made by changing cooking methods and transferring from traditional methods to cooking with a microwave". The saving was difficult to calculate, but Defra suggested an overall saving of 10% might be possible.

**FIGURE 1**: A typical microwave oven has a maximum power of 900 W. About 65% of this power is used to produce microwaves. Some is used to operate lamps, the cooling fan and the turntable. The rest is 'wasted', heating up the surroundings.

## ✳ TASK 1

(a) Microwaves are used in kitchens. Give two more uses of microwaves.

(b) A typical microwave oven has a range of power settings, often from 750 to 900 W. Power is rate of energy transfer:

$$\text{power} = \frac{\text{energy transferred}}{\text{time taken}}$$

If the unit for power is the watt (W) and time is measured in seconds (s), what is the unit for energy in this equation? Give its symbol.

(c) Mains electricity is 230 V. Using:

power = potential difference × current

show that the power of a microwave is about 900 W when the current drawn is 3.9 A.

(d) A 900 W microwave oven takes three minutes at maximum power to heat a pre-cooked meal to eating temperature. How much energy is transferred? Hint: You need to rearrange one of the equations given above. Also, check you use the correct units.

(e) If 1 kW of electricity costs 12p, calculate the cost of using the microwave at full power (900 W) for six hours, defrosting (80 W) for eight hours and medium power (600 W) for 30 minutes.

## ✳ TASK 2

When operating at full power, about 65% of the input energy is transferred to useful energy to produce microwaves and 30% is 'wasted', heating up the surroundings. Draw a Sankey diagram to show the energy transfers that happen in the microwave oven.

## ✴ TASK 3

Here are the results from two tests that compared cooking using a microwave oven and using a conventional electric oven.

**TABLE 1**: Cooking 660g of new potatoes.

| Microwave oven | 600 g new potatoes and 10 g water |
| | Time taken 6 minutes |
| | Energy transferred 0.150 kWh |
| Electric hob | 600 g new potatoes and 1000 g water |
| | Time taken 24 minutes |
| | Energy transferred 0.503 KWh |

**TABLE 2**: Cooking a frozen ready meal.

| Microwave oven | Frozen ready meal for one |
| | Time taken 7½ minutes |
| | Energy transferred 0.177 kWh |
| Electric oven (not pre-heated) | Frozen ready meal for one |
| | Time taken 40 minutes |
| | Energy transferred 0.652 KWh |

(a) Explain the differences in energy transferred by (i) the microwave oven and the electric hob when cooking the potatoes (ii) the microwave oven and the electric oven when heating the frozen meal.

(b) What conclusions can you draw about the use of microwaves ovens for these two cooking activities?

## ✴ TASK 4

(a) In a domestic microwave oven, microwaves have a frequency of $2.45 \times 10^9$ Hz. What do you understand by the term 'frequency'?

(b) Microwaves travel at the speed of light ($3 \times 10^8$ m/s). Calculate the wavelength of the microwaves in a domestic microwave oven to two significant figures. Use the equation: velocity = frequency × wavelength.

(c) Industrial microwave ovens use microwaves with a wavelength of 0.33 m. Calculate their frequency to two significant figures.

(d) The frequencies used in the two microwave ovens is different. What is the difference in energy of the microwaves used. Explain your answer.

(e) Describe how microwave radiation differs from ultraviolet radiation and why it is less dangerous to be exposed to microwaves than to ultraviolet.

## ✴ MAXIMISE YOUR GRADE

| Answer includes showing that you can... |
|---|
| recall uses of microwaves to include mobile phones, satellite TV, cooking. |
| recall the units of energy (joules, J), power (watts, W) and time (seconds, s). |
| recognise and use data to draw and interpret a Sankey diagram. |
| show an awareness of the dangers of using various electromagnetic waves. |
| use and substitute into the equation for power (power = potential difference × current). |
| carry out calculations using power = energy transferred ÷ time. |
| calculate costs of using a microwave using total cost = time × cost per kilowatt-hour. |
| use data to discuss energy use of appliances used in the home. |
| interpret meaning of frequencies as number of waves per second and use the equation: velocity = frequency × wavelength, and show an understanding of significant figures. |

E ⟶ C ⟶ A

*In the examination, equations will be given on a separate equation sheet.*
*Write down the equation that you will use. Show clearly how you work out your answer.*

**1.** For good health, the organs of the body need to work well together.

AO1 **(a)** Name the process which maintains a constant internal environment in the body. [1]

AO1 **(b)** Name the chemical substances which control many of the processes in the body. [1]

**2.** About 2.8 million people in the UK are diagnosed with diabetes. It is estimated that 850 000 people have the condition, but do not know it.

AO2 **(a)** Type 1 diabetes typically develops in children and young adults.

Suggest advice that would be given to a student who had just been diagnosed with this type of diabetes. [2]

AO1 **(b)** Explain how blood glucose levels are maintained in a healthy individual. [3]

AO2 **(c)** Type 2 diabetes usually develops after the age of 40 and accounts for between 85 and 95% of all people with diabetes.

Suggest how this arises and how it can be treated. [2]

**3.** Antacids work by neutralising the acid made in your stomach. Common antacids contain the hydroxides and carbonates of metals. They come under various brand names and in tablets, powders or liquids.

AO1 **(a)** Give the chemical name of stomach acid and describe its action in the stomach. [3]

AO1 **(b)** Give the full name of an active ingredient in an antacid and explain how it neutralises stomach acid. [3]

AO2 **(c)** Hydrogen ions and hydroxide ions react together to form water in a neutralisation reaction. Write a full balanced symbol equation using these ions to show neutralisation. [3]

AO2 **(d)** There are a range of antacid products on the market. A group of students decided to carry out investigative work to find out which were most effective. They decided to carry out a titration and test two antacids. They used a solution of dilute acid. Write down four instructions that the students would need to follow in order to test the antacids. [4]

**4.** Different versions of the same gene are called alleles. The gene for eye colour has an allele for blue and an allele for brown. For any gene, a person may have the same two alleles, or two different ones. Alleles may be recessive or dominant.

AO2 **(a)** What do the terms 'dominant' and 'recessive' mean? [2]

AO2 **(b)** What term is used to describe the condition where, for any one gene, a person has the same two alleles? [1]

AO2 **(c)** What term is used to describe the condition where, for any one gene, a person has two different alleles? [1]

AO1 **(d)** Name two genetically inherited disorders. [2]

**5.** 'Lime' is a term commonly applied to a range of related materials. Quicklime ($CaO$) is formed by strongly heating a form of calcium carbonate such as limestone. Carbon dioxide ($CO_2$) is released and quicklime is left behind.

AO1 **(a)** Write a word equation for the decomposition of limestone to form quicklime using the common names. [2]

AO1 **(b)** Using the information above. Write a balanced symbol equation to show the reactants and products for the decomposition of calcium carbonate. [2]

AO1 **(c)** Quicklime can be mixed with water to form slaked lime, which is called calcium hydroxide $Ca(OH)_2$. Write a balanced symbol equation to show the formation of slaked lime. [2]

**6.** Thousands of years ago humans did not have the technology to tap into the energy resources. However, today a worldwide environmental problem exists due to the effect of the use of non-renewable resources. Over 65% of the world's electrical energy used today is generated by steam turbine generators burning fossil fuels as their source of energy. Combustion of these fuels releases unpleasant gases and solids into the atmosphere.

AO1 **(a)** Explain the term 'non-renewable resources' with suitable examples. [2]

AO2 **(b)** Discuss the problems which can occur to air quality and the effect on global warming of the burning of fossil fuels. [3]

**7.** Electricity is generated at a power station. The electricity is then transferred from the power station to consumers through cables via the National Grid.

AO2 **(a)** Give a reason why electricity transmitted through the National Grid is at a low current. [1]

AO2 **(b)** Power stations produce electricity at 25 000 V but it is sent through the National Grid cables at voltages up to 400 000 V. The voltage of household electricity is about 230 V.

Explain why electricity is sent through the National Grid using such high voltages. [2]

AO1 **(c)** State what is used to reduce the high transmission voltages to safe values that can be used in the home. [1]

**8.** Knowledge of the properties of materials used in construction is important when deciding on the most appropriate material to use. Metals can be divided into ferrous and non-ferrous metals. The use of ferrous material is vast. It is, however, important that the correct material is chosen.

Cast iron tends to be brittle and with a relatively low melting point. It can form many complex shapes made by casting. It has a hard skin but it corrodes.

Mild steel is the most common form of steel because its price is relatively low. It is tough, ductile and malleable. It has good tensile strength but still poor resistance to corrosion.

Medium carbon steel is strong, hard and tough but less ductile than mild steel.

Use the information above and your knowledge and understanding, to answer the questions below.

AO2 **(a)** Give the difference between ferrous and non-ferrous metals. [1]

AO1 **(b)** List three characteristic properties of metals which are useful when using ferrous metals in construction. [3]

AO3 **(c)** Cast Iron can be used in making decorative gates and fences. Why is this considered to be a suitable material? [2]

AO2 **(d)** Medium carbon steel is used in building. What two properties need to be considered. Support your answer with reasons. [3]

**9.** The table below shows the electrical consumption of some household electrical appliances.

| Appliance | Power (watts) | Average usage (hours /day) |
|---|---|---|
| two televisions | 600 | 5 |
| DVD | 200 | 1 |
| audio equipment | 100 | 1 |
| video games | 50 | 1 |
| desktop computer | 300 | 3 |
| laptop | 100 | 2 |
| printer | 200 | 0.1 |
| broadband equipment | 10 | 24 |

AO1 **(a)** State an equation that you can use to calculate the daily power consumption (kWh/day) of the appliances in the table. [2]

AO3 **(b)** Using data from the table, compare the consumption of electricity per day of student A, working on his computer and listening to his music, with student B watching the TV, a DVD and playing video games. [4]

AO2 **(c)** If the cost per unit of electricity is 15p, calculate the cost of electricity spent by student A in a day. [3]

AO3 **(d)** The graph below shows power demand of a household monitored every two minutes. Describe, with reasons, the changes that occur during the day. [4]

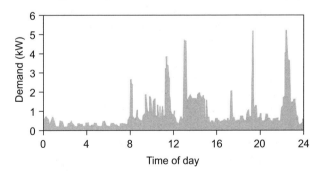

# ✳ WORKED EXAMPLE – Foundation tier

**1.** Within the last 30 years, nuclear energy has become increasingly important in generating electricity. A nuclear power station relies on a controlled chain reaction among atoms of uranium.

Use the information above and your own knowledge to answer the following questions.

**(a)** Give the fuel used in a nuclear power station. [1]

*A radioactive source*

**(b)** Name the type of reaction that allows uranium to produce energy. [1]

*Radioactive decay*

**(c)** Produce four slides to show how electricity can be generated from nuclear fuels. Use the words given to help you. [4]

| Slide number | Words to help you |
|---|---|
| 1 | nuclear reactor |
| 2 | formation of steam |
| 3 | turbine |
| 4 | electricity |

*Slide 1. In a nuclear reactor a reaction occurs to produce heat*

*Slide 2. The heat produced changes water into steam*

*Slide 3. The steam produced then drives a turbine. Energy is transferred to turn the turbine*

*Slide 4. Energy is transferred to electricity*

## How to raise your grade!

Take note of these comments – they will help you to raise your grade.

This answer is insufficient and would receive no mark – remember to read the information given in the question.

The fuel for a nuclear reactor is made from metal called uranium. Uranium undergoes radioactive decay.

This answer is also insufficient and would receive no mark. The candidate needs to know that nuclear fuels produce energy from nuclear fission.

The candidate needs to add more detail to the answer – again use information given in the question. Indicate that in the nuclear reactor the uranium atoms are split to produce a controlled nuclear reaction, which produces heat.

This is fine – a statement that this occurs in the boiler could support the information for this slide.

Information answers the question. The energy transfer information is correct and supports the answer.

Additional detail is needed – that the turbine drives the generator to produce electricity.

# ✳ WORKED EXAMPLE – Higher tier

**1.** The energy efficiency of a lamp can be calculated by using the equation

$$\text{efficiency} = \frac{\text{useful energy out}}{\text{total energy in}}$$

This diagram compares the efficiencies of two lamps:

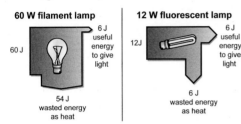

**(a)** Use the equation and the diagram to compare the efficiency of the two lamps. [4]

*The diagram is a Sankey diagram. For the 60 watt filament lamp, only 6 joule energy transferred as light.*
*Efficiency is 6 ÷ 60 × 100 = 10%*
*For the 12 watt fluorescent lamp the same amount of energy transferred as light.*
*However, less energy put in, so more efficient.*
*Efficiency 6 ÷ 12 × 100 = 50%*

**(b)** Suggest two advantages of using fluorescent lamps. [2]

*The light output for the fluorescent lamp is the same as the other. The heat losses for the fluorescent lamp are less.*

**(c)** In this question you will be assessed on using good English, organising information clearly and using specialist terms where appropriate.

Energy-saving tips are widely advertised. How can the claims be checked? Devise an experiment to test materials used to insulate hot water tanks. You are provided with: 100 cm³ beakers, 150 cm³ measuring cylinders, thermometers, timers and a selection of insulating material. [6]

*1. Choose two of the insulation materials and wrap them around two of the beakers.*

*2. Set up three beakers – the two insulated ones and one without any insulation.*

*3. Pour the same amount of hot water in each of them but don't fill them to the top.*

*4. Measure the starting temperature in the beaker. Check that the starting temperature is the same in all the beakers.*

*5. Measure and record the temperature in each beaker every 5 or 10 minutes.*

### How to raise your grade!
Take note of these comments – they will help you to raise your grade.

The candidate has made a good attempt at the answer. It could have been improved if they explained that a Sankey diagram shows energy transfers. The efficiency calculations are correct, but some explanation of the working might have been given.

The candidate might have mentioned that the Sankey diagram expresses energy in joules; this is the amount of energy transferred per second.

It would be better if the candidate separated key points, rather than continuous prose.

The first part of this answer does not state the advantage. A better answer would include data. By using fluorescent rather than filament lamps, the power consumption of the lamps can be reduced from 60 watts to 12 watts for the same useful energy output.

Overall, the information gives a satisfactory outline for the experiment, with information well organised. However, the candidate could have improved the answer by:

> giving an actual volume rather than 'the same amount'

> saying that the thickness of the insulation should be measured

> saying when to stop measuring the temperature

> indicating how the results will be analysed and evaluated, such as graphs to compare rate of heat loss in each of the three beakers.

## What you should know

### Drugs

Drugs alter the way that the body works.

Some drugs affect the brain and nervous system.

Tobacco and alcohol cause long-term damage to body systems.

● Which diseases are linked to alcohol and tobacco?

### Disease

Certain viruses and bacteria can cause disease.

Bacteria can produce poisons called toxins.

Certain white blood cells can make antibodies.

Vaccination can be used to provide immunity to a disease.

● What is a vaccine?

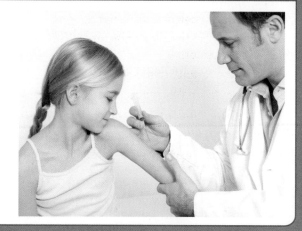

### Atoms

Atoms have a nucleus made of neutrons and protons which is surrounded by electrons.

● Which atomic particles have a positive charge?

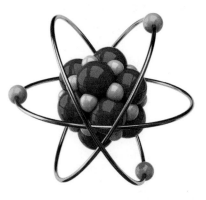

# You will find out

## The use (and misuse) of drugs

> Medicines contain useful drugs that can be used to treat disease.

> Some drugs relieve symptoms but do not cure disease.

> Bacteria but not viruses can be destroyed by antibiotics.

> Over prescription can lead to increased bacterial resistance to antibiotics.

> Drugs must be tested on animals and then humans before they can be put into general use.

> Legal and illegal drugs may be used for recreational purposes.

> Use of drugs may lead to addiction and dependency.

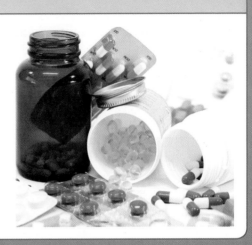

## The use of vaccines

> Viruses cause cell damage.

> Platelets cause the blood to clot to form a barrier to infection.

> Phagocytes engulf pathogens.

> Lymphocytes make antibodies.

> Vaccination is a simple and cheap method to prevent infection.

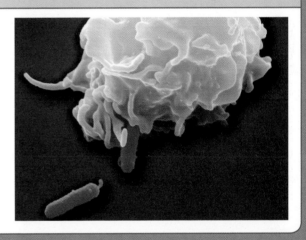

## The use of ionising radiation in medicine

> Radioisotopes emit radiation.

> Radiation can be alpha or beta particles or gamma rays.

> Each type of radiation has characteristic penetrating power.

> X-rays and gamma rays are both useful and harmful.

# Treating disease

**THE SCIENCE IN CONTEXT** With over 1.3 million staff, the National Health Service (NHS) is the largest employer in Europe. It employs nurses, doctors, physiotherapists, clinical scientists and a vast range of other health professionals. These people diagnose and treat disease and illness. To do their jobs they need to understand how a healthy body works and how it becomes unhealthy.

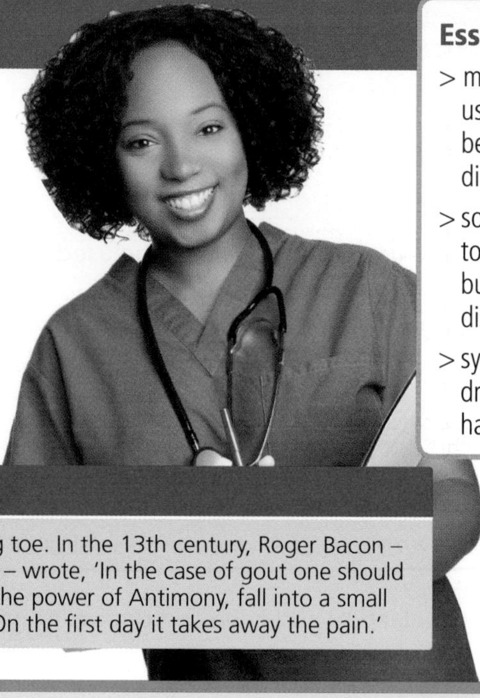

**FIGURE 1**: The NHS employs more than 1.3 million people.

### Essential notes

- > medicines containing useful drugs can be used to treat diseases
- > some medicines help to relieve symptoms but do not cure disease
- > symptom-relieving drugs may cause harm if overused

### Gout medicine

Gout can cause bad pain in joints, especially the big toe. In the 13th century, Roger Bacon – one of the first people to use experimental method – wrote, 'In the case of gout one should let three drops of this Spiriti vini, that has received the power of Antimony, fall into a small glass of wine. This has to be taken by the patient. On the first day it takes away the pain.'

## Drugs and medicines

A **drug** is a substance that is introduced from *outside* the body and alters the normal chemical processes *inside* the body.

Drugs are used for many different purposes. If misused they can cause harm. Some drugs, such as cannabis or heroin, are illegal. Others, such as alcohol and tobacco, have age restrictions.

It is likely that you have taken drugs without even realising it. Many drugs are legal and are used to improve quality of life. A cup of tea or coffee, for example, can improve a persons' mood or make them more alert.

Usually, people think that drugs are there to help them when they are unwell. These types of drug are found in **medicines**. They treat or prevent disease. If you have ever had a headache or suffer from hay fever, you have probably taken a medicine to treat pain or prevent sneezing.

**Remember**
Medicines contain useful drugs that are used in treating disease.

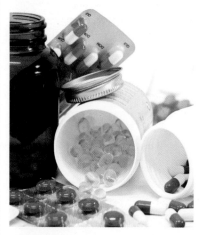

**FIGURE 2**: There are many different types of medicines.

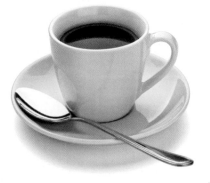

**FIGURE 3**: Coffee contains the drug caffeine. It is one of the most widely sold drugs in the world.

### QUESTIONS

**1** (a) What is a drug?
(b) When is a drug a medicine?
(c) Is insulin a drug or a medicine?

**2** Is coffee a food or a drug?

**3** Why do alcohol and tobacco have age restrictions?

# Types of treatment

Many drugs are used to **cure** or **prevent** disease. For example, antibiotics, such as penicillin, can be used to kill **bacteria**. They can cure diseases such as bacterial meningitis, which affects the membranes (meninges) around the brain. They can also prevent infection in those caring for sufferers.

Other drugs, such as painkillers or anti-inflammatories, may just relieve symptoms without curing their causes.

### Treating symptoms alone

If a **drug** only *masks* symptoms, the disease itself may become worse or the user may become dependent on the drug. Examples of drugs which treat symptoms but do not cure the cause include:

> painkillers, such as aspirin and paracetamol

> treatments for high blood pressure

> antidepressants

> sleeping tablets.

Painkillers are useful to control pain. However, any cause such as a tumour must also be treated.

High blood pressure, depression and insomnia (lack of sleep) may be caused by a stressful life-style. In such cases, stress management and relaxation techniques can help treat the cause of the problem when drugs will not.

**FIGURE 4**: Does the medicine cure the disease or just treat the symptom?

## QUESTIONS

**4** Some drugs only treat symptoms. Are they useful?

**5** Sometimes, treatment by a drug alone is not sufficient to cure a problem. Describe an example of this.

# Drug tolerance

If misused, even legal drugs can cause harm.

Symptom-relieving medicines should only be used for a short period of time. Body chemistry adjusts to the long-term use of drugs.

Drug tolerance occurs when higher doses of a drug are needed to produce the same effect. When the user stops taking the drug, they develop withdrawal symptoms – and then dependency. For example:

> taking painkillers frequently, to control headaches, reduces their effectiveness (drug tolerance)

> it causes further headaches if the user stops taking them (withdrawal symptoms)

> this makes the user dependent on the drug – they need to take the drug even if the original illness is cured.

### Did you know?

In 2007, Linda Docherty died from kidney failure caused by dependency on ibuprofen. She was taking up to 48 tablets a day. There may be as many as 50 000 people in the UK that are addicted to painkillers.

## QUESTIONS

**6** Explain why symptom-relieving medicines should only be used for a short time.

# Bacteria and antibiotics

**THE SCIENCE IN CONTEXT** Treatments of diseases and illness may be medical or surgical. Medical treatments rely on drugs. Some types of drugs, such as antibiotics, can cure a disease. Other types, such as painkillers, only mask the symptoms. In a recent survey of medical experts, antibiotics were voted the second most important medical advance in the past 150 years.

## Penicillin

Penicillin – an antibiotic substance discovered by Sir Alexander Fleming – is obtained from a strain of fungus that needs plenty of oxygen to grow. It is grown in small fermentation tanks of about 40 to 200 dm³. It is estimated that penicillin has saved the lives of at least 200 million people.

**FIGURE 1**: The mould *Penicillium* growing on a culture plate.

## Antibiotics

The first antibiotics were chemicals produced by microorganisms to compete with each other. Antibiotics:

> kill bacteria, or

> inhibit the growth of bacteria.

Penicillin is extracted from a strain of a blue-green **fungus** called *Penicillium*. Relatives of this strain sometimes grow on mouldy bread or fruit.

More recently, artificial alternatives have been made. Chemicals with antibiotic properties have been extracted from other sources, including lichens, toad skin and snowdrops.

**Remember**

An antibiotic is a chemical that can kill or inhibit the growth of most bacteria.

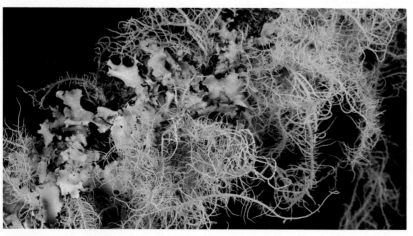

**FIGURE 2**: Usnic acid is a natural antibiotic from lichen.

**QUESTIONS**

1 Explain how antibiotics help to cure infections.

Q … gcse antibiotics

# How antibiotics work

Antibiotics interfere with normal chemical processes only in bacterial cells. These are different from cells found in humans and other animals. So, normally, antibiotics are safe to use.

> Killers: Penicillin prevents bacteria from forming strong cell walls. They take in water by osmosis, swell up and burst.

> Inhibitors: Tetracycline and chloramphenicol inhibit bacterial growth by stopping the bacteria from making proteins.

## Bacterial resistance

Drug companies constantly look out for new **antibiotic** chemicals. This is because bacteria are able to mutate and evolve new strains very rapidly. Existing antibiotics become less and less useful as new resistant bacteria appear. The problem is that, when you use an antibiotic:

> susceptible bacteria are killed

> resistant bacteria remain

> resistant bacteria have no competitors and grow rapidly.

The resistant bacteria replace the bacteria that the antibiotic could kill. This is an example of **natural selection** in action.

## Prescribing antibiotics

A number of factors have caused problems with the overuse of antibiotics.

> When people feel ill they often go to the doctor and ask for an antibiotic.

> Some doctors prescribed antibiotics for viral illnesses like colds and flu.

> In some countries, it is possible to obtain antibiotics without a prescription.

A campaign in Scotland tried to persuade doctors to stop giving antibiotics to everyone that wanted them.

Table 1 shows the effectiveness of this campaign.

| Year | All antibiotics | Penicillin |
|------|----------------|------------|
| 1992 | 95.6 | 51.4 |
| 1995 | 105.6 | 59.8 |
| 1999 | 86.1 | 47.5 |
| 2002 | 82.0 | 44.9 |
| 2003 | 81.9 | 44.7 |
| 2004 | 79.2 | 41.7 |

**TABLE 1**: Number of prescriptions per 100 patients, before and during the campaign.

## Remember
Antibiotics cannot be used to treat diseases caused by viruses.

## QUESTIONS

**2** Use Table 1 to draw a graph. (a) When do you think the campaign started? (b) Is the trend for all antibiotics and penicillin the same? (c) Does the data for 2004 show that 20.8 in every 100 people did not receive any antibiotics? (d) Do you think the campaign was successful? (e) Describe how over-prescription of antibiotics might affect the running of the NHS.

**3** Explain why drug companies try to make new antibiotics.

# MRSA

*Staphylococcus aureus* is a common bacterium found on human skin. Normally it does no harm. Weaker individuals, such as the elderly or patients in hospital, are sometimes unable to combat the bacteria when they enter the body. The bacteria can enter the body, for example, in the lungs or in wounds.

**MRSA** is short for **methicillin-resistant *Staphylococcus aureus***. This strain has been called a 'super-bug' because it is resistant to most antibiotics. It can grow rapidly, causing pneumonia or destroying tissues.

## QUESTIONS

**4** Explain why MRSA has become common.

# Testing drugs

**THE SCIENCE IN CONTEXT** New drugs must be approved by the Medicines and Healthcare Regulatory Agency (MHRA) before they can be put on the market. Following animal studies, an extensive series of clinical trials using human volunteers are carried out. It takes several years before a dossier is submitted to the regulatory authorities. The dossier often consists of tens of thousands of pages of data.

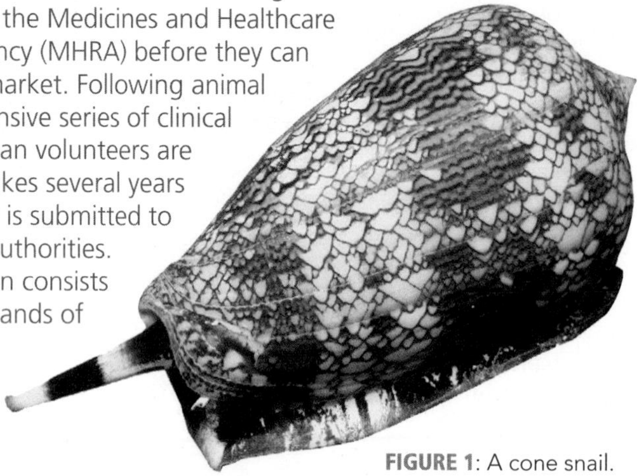

**FIGURE 1**: A cone snail.

**Essential notes**

> drugs must be safely tested on animals and humans before they can be put into general use

**Painkilling snail poison**

Ziconotide is a painkiller made from cone snail venom. Found in tropical coral reefs, the larger snails hunt fish. Their sting has even killed humans. There are more than 600 species and each can produce more than 100 unique poisons – every one is a potential new drug, making cone snails a better source of new medicines than any other group of organisms.

## Testing drugs

Drugs change chemical processes in your body. They can be useful in the treatment of disease, but can also cause harm. It is very important that drugs are thoroughly tested before they are used as medicines.

### Toxicity studies

Toxicity studies establish how poisonous a new drug may be. This involves:

> knowledge of the chemical structure

> testing in laboratories using human cells and tissues grown in cultures

> testing on animals to establish the maximum safe dose that can be used.

Testing new medicines on animals is strictly controlled.

### Clinical trials

The drug is given to healthy human volunteers and then to patients. These trials check:

> Is it safe for humans? The drug should not cause unexpected side effects

> Does it work? The drug must be effective in treating a disease or illness

> What dosage should be used? The correct amount to use to be effective without causing harm.

**Remember**
It takes 10–15 years to discover, develop and test a new drug.

In the UK, the MHRA (Medicines and Healthcare products Regulatory Agency) makes sure that all the necessary tests have been carried out before a new medicine can be marketed. The drug's safety continues to be monitored throughout its use.

Drug trials are very expensive. Each medicine takes years to develop and most drugs are found to be unsuitable.

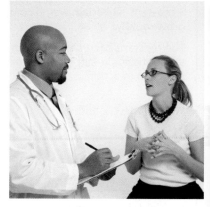

**FIGURE 2**: Doctors must make sure that patients fully understand that they are part of an experiment, before they agree to take part in a drug trial.

**QUESTIONS**

1 Explain why animals, rather than humans, are used in toxicity tests for drugs.

2 Explain why there are strict regulations to control the use of animals for testing drugs.

3 Explain why it is necessary to test new drugs on humans after the animal tests have been carried out.

# Problems with medicines

Sometimes, even after all the testing of a new drug, unexpected side effects occur.

A drug called **thalidomide** was used from 1957 until it was withdrawn in 1961. It was tested as a sleeping pill, but was found to be effective in preventing morning sickness in pregnant women. Around 10 to 20 thousand babies suffered from birth defects, including very short or missing arms. Today, thalidomide is used in the treatment of leprosy and bone cancer.

In 1997, the anti-obesity medicine Fen-Phen was withdrawn after 24 years because it was found to cause **heart disease**.

Cerivastatin, used from the 1990s to prevent heart disease, was withdrawn in 2001 because it caused muscle breakdown, leading to kidney failure.

**FIGURE 3**: A child affected by thalidomide.

### Did you know?

Three-quarters of drugs under development fail to get to market. Drug companies have to spend about £600 million before they find a successful new drug.

## QUESTIONS

**4** Why was thalidomide not tested on pregnant women?

**5** Why has thalidomide not been completely banned?

# Aspirin and personalised medicine

## Aspirin

Willow bark was used for centuries to treat fevers and pain, before the discovery of its active ingredient, salicylic acid. This was modified to give aspirin. It is not known how many deaths have been caused by the overuse of aspirin. The doses used to treat flu in the great **pandemic** of 1918 were at toxic levels and probably killed many patients who might have recovered.

Reye's syndrome is a potentially fatal disease that can develop in children who are given aspirin when suffering from a viral disease. It affects many organs, especially the brain and liver. Since aspirin-taking by children has been stopped, there have been many fewer cases of Reye's.

**FIGURE 4**: Willow bark was used to treat fevers and pain.

## Personalisation of medicines

Modern approaches to treating disease include taking genetic and other factors into account when prescribing drugs. This is to make prescriptions as safe and as effective as possible for the individual.

## QUESTIONS

**6** An important rule in medical ethics is 'first do no harm' (from the Latin motto: *primum non nocere*). Discuss with others: Does the use of drugs follow this rule?

 ... aspirin molecule

# Legal and illegal drugs

**THE SCIENCE IN CONTEXT** Most drugs are legal. Some require a doctor's prescription, others do not. All come with advice on their use, including recommended dose and possible side effects. Some drugs are illegal and used for recreational purposes. Performance-enhancing drugs (legal or illegal) are banned in most sports. The laboratory to test athletes during the 2012 Olympic and Paralympic Games cost £10 million.

FIGURE 1: Opium poppies are used to make heroin. Heroin and cocaine are seen as the two most dangerous recreational drugs.

## Illegal drugs use

The illegal use of drugs causes wide-ranging serious problems. Drug addiction ruins lives. It can affect children as well as adults. Addiction forces people into crime and poverty. Drug dealing brings crime and violence into communities.

## Legal or illegal?

Certain drugs such as alcohol, tobacco or prescription drugs are legal if used in the correct circumstances. They may become illegal if used, for example, by someone who is under age or who has not been prescribed the drug by a doctor. Some drugs, like heroin and cocaine, are banned by law.

### Recreational drugs

Recreational drugs are taken to alter mood, to give a lift or make someone feel more relaxed or happy. They may be legal, such as alcohol and tobacco, or they may be illegal, such as heroin and cocaine.

### Types of drugs

There are five main types of drug.

| Type | Examples | Effects |
|------|----------|---------|
| stimulant | nicotine, caffeine, ecstasy, amphetamines, cocaine | raise brain activity, increase alertness |
| depressant | alcohol, barbiturates, solvents | slow down brain activity and reactions |
| hallucinogen | cannabis, LSD | distort vision and hearing |
| painkiller | heroin, morphine | block nerve impulses, can cause deep relaxation and euphoria |
| performance enhancer | anabolic steroids | increase muscle development |

**Did you know?**

Possession of Class A drugs such as heroin or ecstasy can lead to a seven-year prison sentence.

## QUESTIONS

1 When can a legal drug be illegal?

2 Describe the difference between a medicinal drug and a recreational drug.

3 Suggest why some employers carry out drug testing in work places.

# Addiction and dependency

Because they change chemical processes in the body, all drugs have the potential to cause harm and damage health. The body adapts to the changes, leading to complex physical and psychological symptoms.

**Addiction** and dependency are closely related. A person becomes addicted when they are unable to stop using a drug. There is a shift to dependency when they 'can't live without the drug'.

Alcohol, tobacco, heroin and cocaine are all addictive. Body and brain chemistry change so that unpleasant withdrawal symptoms occur if the person stops taking them. They then have to take the drug to prevent themselves from feeling very ill.

## The effects of addiction

When the body adapts to a drug, it has several effects.

> More of the drug is needed to give the same effect.

> The person feels unwell when the effects of the drug wear off.

> There is a persistent strong desire to take the drug.

> Social activities are reduced.

> Drug use continues despite any physical or mental problems.

FIGURE 2: Cannabis was once thought to be relatively harmless, but now doctors believe that it may cause mental illness.

## QUESTIONS

**4** How can drugs cause addiction and dependency?

# Drug abuse in sport

Athletes are banned from taking many drugs that might enhance their performance and give them an unfair advantage. These include:

> anabolic steroids and certain **hormones** – to help build muscle

> strong painkillers – to overcome the pain barrier

> stimulants – to raise heart rate and improve performance

> beta-blockers – to increase steadiness for precision sports.

Drug abuse may also cause long-term damage to the heart, liver and brain and drastically reduce life expectancy.

Overdose can also kill. 'Scotland's strongest woman', the Scottish bodybuilding champion, Louise Halliwell, died at the age of 38, owing to an **insulin** overdose that caused massive brain damage.

The World Anti-Doping Agency (WADA), lists the banned substances and coordinates the international fight against drug abuse in sport.

## QUESTIONS

**5** Explain why some athletes use banned drugs.

**6** Some people believe that athletes should be able to use any of the drugs that are currently legal outside sport to improve their performance. What do you think?

FIGURE 3: Some professional cyclists have taken EPO, a drug that increases the number of red blood cells. How might this help their performance?

# Tobacco and alcohol

**THE SCIENCE IN CONTEXT** Tobacco and alcohol are legal recreational drugs. The effects of these legal drugs on the human body have been studied by medical professionals and scientists. They have found reliable evidence that links the misuse of tobacco and alcohol to breathing (respiratory) disorders and to circulatory (heart and blood vessels) disorders.

## Egyptian beer

Beer was a very important drink in ancient Egypt, for adults and children alike. Scenes of beer and wine brewing are depicted on the walls of tombs. The drinks were fermented in large vats and decanted into large jars for storage or transport.

**FIGURE 1:** Ancient Egyptians brewed beer and wine on a large scale.

## Dangerous pleasures

Although tobacco and alcohol are legal recreational drugs, they are among the most prevalent causes of preventable deaths.

Tobacco smoke contains about 4000 different chemicals, of which many are harmful to the body. They damage the heart, lungs and blood vessels. It is the nicotine which is highly addictive, causing smokers to develop a high dependency on tobacco.

Alcohol in alcoholic drinks such as wines, beers and spirits, is a poison. It causes brain and liver damage. It is a depressant that slows down the nervous system. Regular drinking becomes addictive, leading to dependency.

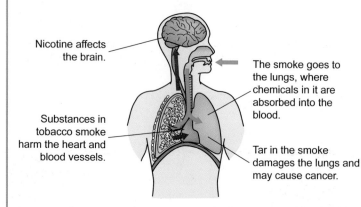

Nicotine affects the brain.

The smoke goes to the lungs, where chemicals in it are absorbed into the blood.

Substances in tobacco smoke harm the heart and blood vessels.

Tar in the smoke damages the lungs and may cause cancer.

**FIGURE 2:** How tobacco smoke affects organs in the body.

### QUESTIONS

1 Should tobacco and alcohol be legal? Give a reason for your answer.

2 Why is it difficult to give up smoking, once you have started?

Q ... gcse tobacco

# The dangers of tobacco and alcohol

## Tobacco

Substances in tobacco damage the respiratory and circulatory systems.

> Cilia are killed so mucus accumulates in the lungs, causing infections such as bronchitis.

> The lungs are damaged, causing COPD (chronic obstructive pulmonary disease).

> Chemicals which are carcinogenic (especially in tar) cause lung cancer.

> Damage to blood vessels increases the risk of heart attacks and strokes.

Carbon monoxide in the smoke reduces the oxygen-carrying capacity of the blood. It combines permanently with haemoglobin, the oxygen-carrying **molecule** in red blood cells.

## Alcohol

Alcohol is a depressant.

> It increases reaction times and impairs judgement.

> A small amount is relaxing but larger quantities lead to a loss of self-control.

> Alcohol drinkers have a much higher risk of accidents.

> Excessive intake can cause coma and death.

One in three people attending Accident and Emergency Departments have been drinking. About 15% of road accident fatalities are caused by alcohol.

FIGURE 3: If you drink and drive, you increase the chance of killing and being killed.

### QUESTIONS

**3** In what ways are the effects of tobacco and alcohol (a) similar (b) different?

**4** Suggest what effects a reduction in oxygen-carrying capacity of the blood may have in the body.

# Deaths caused by tobacco and alcohol

People are often unaware of the dangers of tobacco and alcohol, because they are a part of normal life in the UK. Nonetheless, health authorities and the Government are trying to reduce consumption to lessen the harmful effects on health and the high cost to the National Health Service.

| Year | 2000 | 2001 | 2002 | 2003 | 2004 | 2005 | 2006 | 2007 | 2008 |
|------|------|------|------|------|------|------|------|------|------|
| Men | 4483 | 4938 | 5069 | 5443 | 5431 | 5566 | 5768 | 5732 | 5999 |
| Women | 2401 | 2561 | 2632 | 2721 | 2790 | 2820 | 2990 | 2992 | 3032 |

TABLE 1: The number of deaths from drinking alcohol, in each year from 2000 to 2008. These data are for England and Wales.

### QUESTIONS

**5** Explain the trends (patterns) shown by the data in Table 1.

**6** (a) Explain why so many young people want to smoke and drink. (b) Do you think that advertising bans or restricted sales will reduce tobacco and alcohol consumption? (c) If you could, would you ban tobacco and alcohol from sale in the UK? Explain your answer.

# Preparing for assessment: Analysing data and evaluating the investigatio

*To achieve a good grade in science, you not only have to know and understand scientific ideas, but you need to be able to apply them to other situations and investigations.*

Connections: This task relates to Unit 1 context 3.5.1.1 The use (and misuse) of drugs. Working through this task will help you to prepare for the practical investigation that you need to carry out for your Controlled Assessment.

## Investigating the effect of caffeine on reaction time

Jay had read that drinking caffeine makes people react faster. She decided to investigate this. She found six friends to help her and set up a trial.

Jay found a website where you clicked a button to start and waited for a traffic light to turn green. As soon as it did, you clicked the button again. The website measured how long it took between the appearance of the green light and you clicking the button. It gave the time to the nearest millisecond (ms).

She put six drinks, labelled A–L, on a table and let her friends choose one each. Three of them (A–C) were ordinary coffee (containing caffeine), and three (D–F) were decaffeinated coffee (containing little or no caffeine). However, she simply told her friends that all the drinks were coffee. Her friends randomly chose a drink. This is called a blind trial.

Jay asked each of her friends to do the reaction time test five times (leaving just a few seconds between tests), then drink a mug of coffee (not knowing whether it was ordinary coffee or decaffeinated) and after 15 minutes do the test another five times.

| | **Reaction time (in milliseconds, ms)** | | | | | |
|---|---|---|---|---|---|---|
| **Person** | **Before drinking coffee** | | | | | |
| A | 224 | 221 | 215 | 209 | 208 | 210 |
| B | 288 | 280 | 277 | 278 | 276 | 274 |
| C | 239 | 236 | 228 | 228 | 226 | 225 |
| D | 249 | 249 | 246 | 247 | 245 | 246 |
| E | 278 | 271 | 268 | 266 | 262 | 263 |
| F | 234 | 229 | 228 | 225 | 226 | 228 |
| **Person** | **After drinking coffee** | | | | | |
| A | 207 | 201 | 197 | 195 | 196 | 195 |
| B | 258 | 252 | 253 | 250 | 251 | 250 |
| C | 220 | 219 | 213 | 216 | 212 | 214 |
| D | 250 | 248 | 248 | 249 | 247 | 247 |
| E | 276 | 268 | 254 | 265 | 262 | 261 |
| F | 236 | 235 | 229 | 230 | 228 | 229 |

**TABLE 1**: Results of the blind tests.

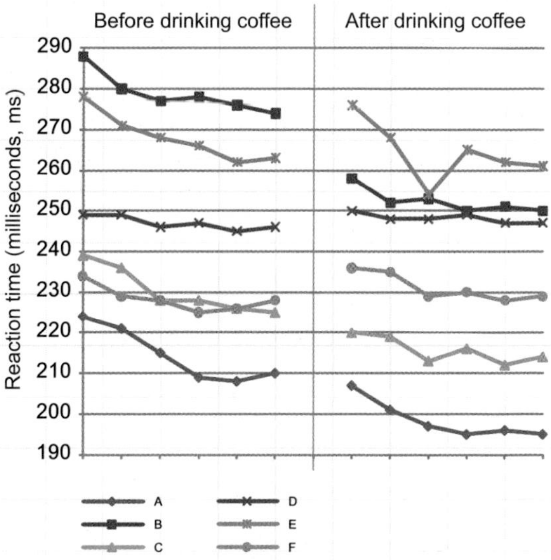


**FIGURE 1**: Graphs to show trends and patterns.

## ✳ Analysing data

**1.** After looking at the results carefully, Jay decided that it was not appropriate to calculate a mean for any of them. Give your reasons for agreeing or disagreeing with her.

> If you look carefully, you will see a pattern in the results. Think about what happens when you practise doing something.

**2.** Jay decided that, although at first sight her results suggested that caffeine reduces reaction time, she could not confidently conclude that this was so. Give as many reasons as you can why she was right to be cautious.

> Your answer to question 1 might help here.

**3.** Describe how you would find (a) secondary data to compare your results with (b) information that will help you give a scientific explanation for the results.

> Remember that a single source of information is not likely to be sufficient. Also, you cannot be confident that the reliability of information from all sources is the same.

**4.** Jay found a report on the internet of a similar experiment. The investigator had used the same reaction time test, but in a slightly different way. During one week, on each day he drank half a cup of coffee and, after 7½ minutes, did the test again. Then he drank the remaining coffee and waited a further 7½ minutes before doing the test again. Here are his results:

| Reaction time (milliseconds) | Mon | Tues | Wed | Thurs | Fri |
|---|---|---|---|---|---|
| before drinking coffee | 408 | 308 | 300 | 291 | 355 |
| after half a cup of coffee | 350 | 306 | 264 | 262 | 292 |
| after drinking all the coffee | 298 | 381 | 305 | 348 | 297 |

Describe how your results compare with this secondary data.

> Think about the differences and the similarities between Jay's investigation and the one she found on the internet.

## ✳ Evaluating the investigation

> Think about variables and the importance of a control.

**5.** Explain why Jay used a mixture of drinks – some containing coffee with caffeine and some containing coffee without caffeine.

**6.** Explain why Jay did not tell her friends which type of coffee they were drinking.

> Do you think that you might pay more attention and concentrate harder on a task if you knew what was being expected? Perhaps find out about a 'placebo'.

**7.** Suggest ways in which Jay could modify her experiment to obtain results that would make her more confident about deciding if caffeine does reduce reaction time.

> You might consider, for example, the sample size or the idea of a 'double blind' trial, and a number of other variables.

# Infection

**THE SCIENCE IN CONTEXT** Infection by harmful bacteria and viruses is a major cause of disease. The bacteria or viruses are passed from person to person – by direct contact or indirectly. Despite all the amazing advances in medicine, in a recent survey of medical experts, sanitation (hygiene and sewage disposal) was voted the most important medical advance in the past 150 years. It is a highly effective way to reduce the spread of infection.

## Bird flu

Doctors and scientists feared that bird flu would spread worldwide very fast and infect a very large number of people. Disease on this scale is called a pandemic. It has not happened because the virus cannot pass easily from one person to another.

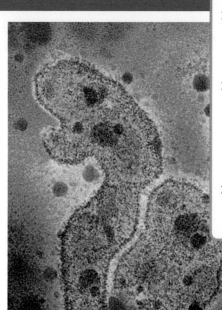

**FIGURE 1**: H5N1 (bird flu) viruses bursting out of a human cell and destroying it.

## Pathogens

Organisms that cause disease are **pathogens**. They include microorganisms such as **bacteria** and **viruses**.

Diseases caused by bacteria include:

> tuberculosis (TB) – bacteria are spread in droplets in the air and infect the lungs

> cholera and typhoid – bacteria are spread in contaminated food and water and infect the gut.

Diseases caused by viruses include:

> measles, mumps, rubella – viruses are spread in droplets in the air, which enter the lungs or are transferred from surfaces touched by the hands

> polio – transmitted in water contaminated with faeces or on unwashed hands after using the toilet.

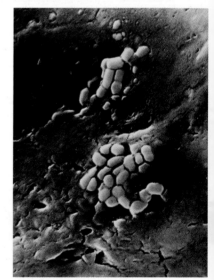

**FIGURE 2**: Bacteria on roast beef (magnified 2000 times).

### Did you know?

Most bacteria are harmless or helpful. They include decomposers essential for breaking down dead organisms.

### Remember

Diseases are caused by bacteria and viruses, which can penetrate the body's defences.

### QUESTIONS

**1** Explain why it is important to (a) cover your mouth and nose when sneezing (b) wash your hands regularly, especially after using the toilet.

Q ... gcse pathogens

# Spreading pathogens

## How pathogens enter the body

The human body is normally good at keeping out pathogens. Skin is an effective waterproof barrier. However, bacteria and viruses are able to enter the body through:

> wounds to the skin – from contaminated surfaces

> the respiratory system – in droplets in the air

> the digestive system – in contaminated water or food

> the genitals – through sexual contact with infected people.

Infected people can transfer pathogens directly – by contact with others – or indirectly via surfaces, food and drink or sneezed into the air.

The cycle of infection can be broken by:

> keeping surfaces clean with disinfectants

> catching sneezes in tissues, which are then disposed of hygienically

> regular hand washing with soap and warm water

> avoiding contact with infected people.

## Becoming ill

Warmth, food and moisture provide perfect conditions inside the body for the rapid reproduction of pathogens.

Bacteria grow and divide rapidly to produce **toxins**. These poisons are carried all round the body in the blood. They cause the symptoms of the disease, such as fever, making us feel ill.

Viruses reproduce inside host cells, which they turn into virus factories. The host cell manufactures huge numbers of new viruses. These new viruses burst out of the cell, destroying it, and infect many more new cells. It is not long before a large amount of **cell damage** has been done, causing the symptoms of the viral disease, such as a sore throat.

**Remember**
One 'bacterium' and two or more 'bacteria'.

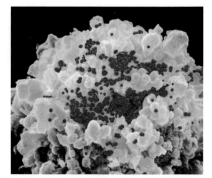

**FIGURE 3**: The round viruses bursting out of the cell are HIV. Contact with this deadly virus can be prevented by using a condom.

## QUESTIONS

**2** List four ways in which infection can occur, naming an example of a disease for each method of transmission.

**3** How do (a) bacteria (b) viruses make us feel ill?

# Health and hygiene

Childbed fever used to kill many women giving birth in hospital. In the 19th century, doctors did not know that disease was caused by microorganisms. Hospitals were not kept clean and doctors did not wash their hands.

Ignaz Semmelweiss, a doctor at a Vienna hospital, noticed that death rates were much lower in the labour ward supervised by midwives than in the ward where doctors were being trained. Then, a professor was accidently cut while carrying out a post-mortem on a victim of childbed fever. He died after developing the same symptoms.

Semmelweiss introduced hand washing with chlorinated lime for doctors before they examined the women in labour. The death rates in the doctors' ward fell dramatically to the same as the other ward.

## QUESTIONS

**4** Explain why the hand washing reduced the incidence of childbed fever.

# The body's defence

**THE SCIENCE IN CONTEXT** Blood is your first line of defence against infection, so it is important to understand it. Haematologists are medical scientists who study blood, looking for abnormalities within the different types of blood cells. Tests they carry out can help to identify diseases, such as anaemia (not enough red blood cells) and leukaemia (too many white blood cells).

## Pus!

Pus contains millions of dead bacteria and white blood cells. A really bad infection requires antibiotics to prevent it spreading through the body.

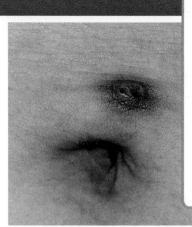

FIGURE 1: Bacteria are thriving in this girl's piercing. It is inflamed – hot and swollen – because her body is sending extra blood to fight them.

## Blood defences

Your blood helps to defend you against infection. It helps to keep pathogens out and attack and kill them if they get in.

> Blood clots form a barrier if the skin is cut.

> Some white blood cells engulf (surround and take in) and kill bacteria.

> Other white blood cells release **antibodies** to destroy bacteria and viruses.

The body has to make the right antibodies to attack a particular pathogen. It takes time to do this. After that the body can make them quickly if the pathogen returns. You are then **immune** to the disease that it causes.

### Did you know?

Tears contain an antiseptic enzyme, lysozyme, which kills bacteria.

### QUESTIONS

**1** What two methods does the blood use to defend the body against infection?

**2** (a) Which blood cells kill bacteria? (b) What two methods do they use?

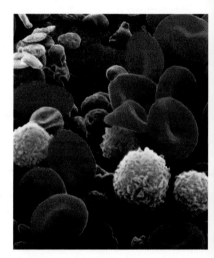

FIGURE 2: Magnified cells from a drop of blood. The white cells have been coloured yellow. In your body, they are actually colourless and transparent.

## Platelets, phagocytes and lymphocytes

### Platelets

There are small particles in the blood called platelets. If you cut your skin, you bleed. The platelets come into contact with the air and trigger reactions which cause the blood to clot. Red blood cells are trapped by fibres and form a solid clot, which plugs the wound.

A blood clot stops you from losing blood, but also forms a barrier to infection. It prevents pathogens from entering into the body through the wound in the skin.

Q ... gcse body defences

## Phagocytes and lymphocytes

There are two kinds of white blood cell:

> **phagocytes** engulf pathogens, such as bacteria

> **lymphocytes** make antibodies.

**Phagocytosis** is the method used by a phagocyte to engulf and digest bacteria. More blood flows to a wound or site of infection, bringing many phagocytes to attack any bacteria or other foreign cells, not normally found in the body. This happens very quickly.

1 A phagocyte moves towards a bacterium.

2 The phagocyte pushes a sleeve of cytoplasm outwards to surround the bacterium.

3 The bacterium is now enclosed in a vacuole inside the cell. Enzymes then kill and digest it.

**FIGURE 3**: Phagocytosis.

**Lymphocytes** defend the body against longer-lasting attacks, and prevent future attacks by the same pathogen. There are many kinds of lymphocytes. Each can make a different antibody. When a pathogen enters the body, it is detected by the lymphocytes that can make the correct antibody to destroy it.

On the first attack, it takes time to make enough antibody to get rid of the pathogen. You may become ill. If the pathogen comes back, antibody is made quickly and it is destroyed before you become ill again – you are **immune** to that disease.

> ### Remember
> An antibody is a chemical made by white cells inside the body that kills one kind of bacterium.
> An antibiotic is a chemical made outside the body that kills most types of bacteria.

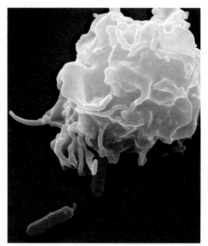

**FIGURE 4**: A phagocyte engulfing bacteria.

### ● QUESTIONS

**3** Describe, in your own words, what a phagocyte does.

**4** Describe, in your own words, what a lymphocyte does.

## Antibodies are fussy

Antibodies are Y-shaped molecules. The ends of the arms have a particular shape. Each pathogen has certain unique molecules (called antigens) with a particular shape.

The correct antibody molecules and the unique pathogen molecules fit together like the pieces of a jigsaw puzzle.

### ● QUESTIONS

**5** Explain why, if the body becomes immune to mumps, it is not immune to measles.

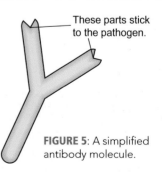

These parts stick to the pathogen.

**FIGURE 5**: A simplified antibody molecule.

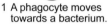

Q … gcse antibodies

# Vaccination

## THE SCIENCE IN CONTEXT

Recovery from some infectious diseases means that you will never get them again. You have become immune to them. By understanding how this happens, medical scientists have been able to develop a vaccination, a technique that prevents certain diseases occurring in the first place. It has saved lives and prevented disability in millions of people worldwide.

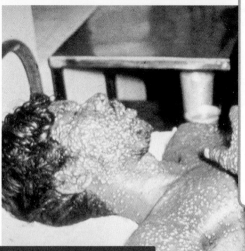

## Defeating disease

The earliest evidence of human smallpox is found in the mummy of the Egyptian Pharaoh, Ramses V, but it has been around for 12 000 years. In the 20th century alone, the virus killed somewhere between 300 and 500 million people, many of them children. By 1980, vaccination campaigns finally eradicated the disease.

FIGURE 1: This Bangladeshi man was one of the last people to get smallpox.

## Immunity

You have learnt that white blood cells that make antibodies are **lymphocytes**. You have many different kinds. Each one is able to make a specific **antibody** that will kill a particular pathogen.

### Becoming immune

A new pathogen enters the body:

> lymphocytes that make the correct antibody multiply

> lymphocytes make antibody and release it into the blood

> there is a delay, pathogens may multiply and make you ill

> the level of antibody rises and the pathogen is destroyed – you get well.

The pathogen enters the body again:

> the correct lymphocytes are ready to make the antibody

> antibody levels rise fast

> the pathogen is destroyed before it can make you ill. You are now immune to that disease.

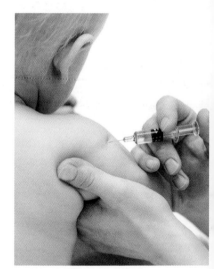

FIGURE 2: Why might injections be better than infections?

## Did you know?

Microsoft billionaire Bill Gates has donated hundreds of millions of dollars towards the campaign to eradicate polio. There has been a 99% reduction in cases over 20 years.

## QUESTIONS

1 Explain how a measles vaccination makes you immune to measles.

Q ... gcse immunity

# Immunisation

Some pathogens cause deadly diseases. If your body is unable to make antibodies quickly enough during the first infection, you may suffer serious harm or even die.

**Immunisation** is used to teach your body how to make antibodies before you are infected by a pathogen. You have probably been immunised against dangerous diseases such as measles, mumps, rubella, polio and diphtheria.

**Vaccines** contain a dead or mild form of a pathogen. Usually, **vaccination** is carried out by injecting a vaccine into a muscle, but some vaccines can be given by mouth. Lymphocytes then make antibodies, just as though the real pathogen was present.

You are now immune to that disease. If you are infected by the pathogen in the future, your body is ready to make the antibodies rapidly. You do not have to suffer from the disease first.

**Did you know?**

A 2010 report showed that 20% of childhood deaths in England and Wales are caused by infections.

## QUESTIONS

**2** What is the most important ingredient in a vaccine?

**3** Why are vaccines used when the body can become immune on its own?

# The importance of vaccination

**MMR** vaccine helps to give **immunity** against measles, mumps and rubella. Before it was first used in 1988, mumps was a common infection. It caused 1200 children to be sent to hospital each year. By 1996, the number of cases fell to less than 100.

Vaccination rates for MMR and the number of mumps cases per year are shown in Table 1. Since 2005, vaccination rates have remained at about 85%. Mumps cases fell dramatically in 2006 and 2007, but have been steadily increasing again to exceed the 2004 levels. To be effective, vaccination rates should be about 95%.

Some vaccines can have side effects and may even cause injury and death. Some experts think that the widespread use of vaccines reduces natural immunity and makes people at greater risk of contracting diseases.

In 1998, a report suggested that the MMR vaccination could increase the risk of autism and certain bowel diseases. This was widely reported in the media. Since then, no evidence has been found to support the link. In 2010, the report was withdrawn as 'dishonest'.

| Year | Percentage of children having MMR vaccination | Number of mumps cases |
|---|---|---|
| 1996 | 92 | 94 |
| 1997 | 92 | 180 |
| 1998 | 91 | 119 |
| 1999 | 88 | 372 |
| 2000 | 88 | 703 |
| 2001 | 87 | 777 |
| 2002 | 84 | 502 |
| 2003 | 82 | 1549 |
| 2004 | 83 | 8104 |
| 2005 | 85 | over 43 000 |

**TABLE 1**: MMR vaccination and cases of mumps.

## QUESTIONS

**4** Which diseases does MMR vaccine protect against?

**5** What effect did the introduction of the MMR vaccine have on cases of mumps?

**6** Why are some people worried about having their children vaccinated?

**7** Suggest reasons for the number of mumps cases in 2005.

**8** Suggest a possible explanation why mumps cases are rising again.

Q ... gcse vaccination ... mmr vaccine

# Nuclear radiation

**THE SCIENCE IN CONTEXT** Physicists have found that some elements emit radiation – they are said to be radioactive. There are three types of emitted radiation. Two are moving particles (slow-moving helium nuclei and fast-moving electrons); the third is waves (electromagnetic radiation). Scientists have found many uses for this radiation.

### Nuclear diagnosis

Nuclear medicine uses radiation to diagnose and to treat illness. About 90% of procedures using nuclear radiation are for diagnosis. The diagnostic tests use substances that emit gamma radiation inside the body.

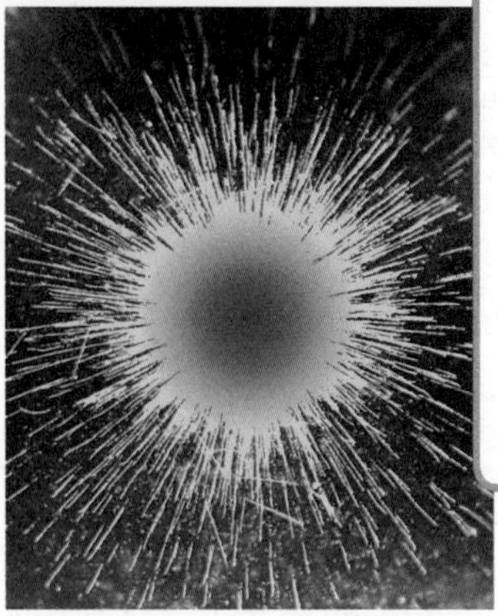

**FIGURE 1**: Alpha particles being emitted by radium.

### Essential notes

- radioisotopes can decay, emitting radiation
- radiation may be alpha or beta particles or gamma rays
- each type of radiation has characteristic penetrating power

## Three types of radiation

Each **element** always has its unique number of protons or electrons. In some elements, the number of neutrons in the **nucleus** varies, giving different isotopes.

Radioactive isotopes (radioisotopes) are unstable because they have extra neutrons. They release nuclear radiation when their nuclei disintegrate. This is radioactive **decay**.

In radioactive decay, the unstable isotope is attempting to become stable.

- The number of protons changes.
- Radiation is emitted.
- The nucleus changes and forms a different sort of element.

There are three types of nuclear radiation: alpha particles, beta particles and gamma rays.

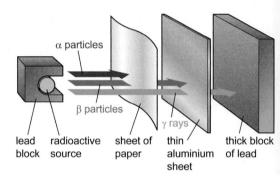

**FIGURE 2**: The three radiations and their penetrating properties.

| Type | Description | Penetrating power |
|---|---|---|
| **alpha particles, α** | > helium nuclei (no electrons)<br>> slow moving (one-tenth speed of light)<br>> large, relatively heavy | > poor<br>> travel short distance in air<br>> stopped by a sheet of paper |
| **beta particles, β** | > electrons<br>> fast moving<br>> small, light | > good<br>> stopped by 3 mm thick aluminium sheet |
| **gamma rays, γ** | > waves of electromagnetic spectrum<br>> travel at speed of light<br>> very high energy<br>> no mass (not particles) | > excellent<br>> 10 mm thick lead stops about half |

**TABLE 1**: Some characteristics of nuclear radiation.

### QUESTIONS

1 Describe what happens to atoms when they release radiation.

2 Which radiation (a) is waves? (b) is electrons? (c) has the lowest penetrating power? (d) travels fastest?



... gcse nuclear radiation

# Ionising radiation

Any of the three types of nuclear radiation can strike an **atom** with sufficient **energy** to dislodge electrons. If an atom loses one or more negatively charged electrons it becomes a positively charged ion.

## Alpha particles

These helium nuclei ($^4_2$He) are heavy positively charged particles. They are good ionisers. As they have poor penetrating **power**, they cause little harm externally to the body. They may cause damage internally, for example, if a source is swallowed.

## Beta particles

These are electrons, so they have a negative **charge**. They are 1/ 7000th of the size of an alpha particle. They move much faster, at about half the **speed of light**. They can collide and **ionise** atoms, but mainly miss them because of their very small size.

Although they are only weak ionisers, they are not safe. Workers using beta particle sources must wear protective clothing.

## Gamma rays

These very high-**frequency** waves have very high energy, but as they have no mass or charge they are very poor ionisers. They are dangerous because their high energy can damage living cells. Careful safety measures are essential when working with gamma rays.

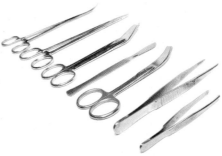

**FIGURE 3**: Surgical instruments used to be heat sterilised. Now they are irradiated with gamma rays to sterilise them.

## QUESTIONS

**3** Which radiation is the best ioniser? Explain your answer.

**4** Do gamma rays have high or low wavelengths?

**5** (a) Which radiation type is most dangerous? (b) Why is it dangerous?

# Half-lives

As the atoms in a sample of radioisotope break down and emit radiation, the number of atoms of the original isotope decreases. There will be a time over which half of the unstable atoms break down and half of them are left.

Each isotope has its own unique rate of radioactive decay. Each has a definite time for half of the unstable atoms to decay which is called its half-life.

It will take the same time for half of the remaining atoms to decay, then again for the next half and so on.

The half-life is always the same and can be used to identify unknown radioactive sources.

| Isotope | Radiation | Half-life |
|---|---|---|
| uranium-238 | alpha 5000 | 4½ billion years |
| plutonium-239 | alpha and beta | 24 000 years |
| carbon-14 | beta | 5700 years |
| cobalt-60 | gamma | 5 years |
| americium-241 | alpha | 460 years |
| iodine-131 | beta | 8 days |
| sodium-24 | beta | 15 hours |
| strontium-93 | beta and gamma | 8 minutes |
| barium-143 | beta | 12 seconds |
| polonium-213 | alpha | 0.000 004 second |

**TABLE 2**: Half-lives of some radioisotopes.

## QUESTIONS

**6** Explain how half-life affects the radioactivity of a material.

# Radiotherapy

**THE SCIENCE IN CONTEXT** Radiotherapy is the use of high-energy nuclear radiation to treat diseases. This radiation is ionising and damages or destroys cells in the body. Cancer cells are targeted and bombarded with gamma rays, making it impossible for them to grow. However, treatment with nuclear radiation has ethical issues that may have to be considered.

## Using scanners

MRI (magnetic resonance imaging) and PET scanning (positron emission tomography) allow radiotherapists to locate tumours exactly inside the body. They target cancers precisely so that they absorb the maximum amount of radiation. Surrounding healthy tissues and organs receive much less radiation and suffer minimal damage.

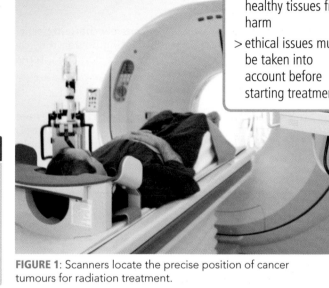

**FIGURE 1**: Scanners locate the precise position of cancer tumours for radiation treatment.

## Radiotherapists

**Radiotherapy** uses high-energy ionising radiation. **Radiotherapists** (also called therapeutic radiographers) are health-care professionals that use radiotherapy to kill cancer cells.

**X-rays** and gamma rays are high frequency **electromagnetic waves**. They can penetrate body tissues and ionise atoms. A fine beam of these high-energy waves can be aimed at a tumour, inside a patient's body, to kill cancer cells.

### Did you know?

Marie Curie pioneered the radiation treatment of cancers. Yet at the beginning of the 20th century, prejudice against women scientists was so great that she was never elected to the French Academy of Sciences.

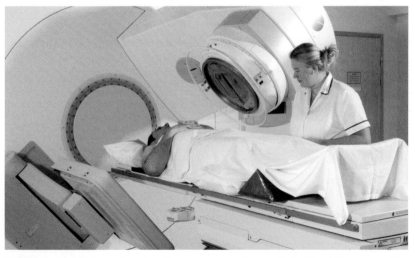

**FIGURE 2**: A radiotherapist using the gamma rays from cobalt-60 to kill cancer cells. This is called cobalt irradiation.

### QUESTIONS

**1** Explain why X-rays and gamma rays are used to treat cancer.

**2** Which of these is radioactive and which is irradiated:
(a) person treated with radiotherapy (b) cobalt-60 (c) cancer tumour exposed to gamma rays?

🔍 … gcse uses of radiation

# Killing cancer cells

Ionising radiation damages **DNA**. Normal cells are able to repair DNA that has not been too badly damaged. Cancer cells are less able to repair the DNA. They are more likely to be killed or unable to continue to grow.

## External radiotherapy

**External radiotherapy** uses an external source of radiation. For example, an X-ray tube or a source of gamma rays.

Radiation travels through healthy tissues to reach a tumour, so radiotherapists take care to reduce the exposure of healthy cells.

> The tumour is located accurately, for example, by using PET (positron emission tomography).

> The patient is positioned accurately. Lasers and moulds are used to keep the patient in the exact position for the radiation to hit the tumour.

> Radiation beams are aimed from different angles. The **absorbed** dose is higher at the tumour than in the surrounding tissues.

> Doses are given in fractions. A small dose is given several times. Healthy cells can recover between treatments.

## Internal radiotherapy

**Internal radiotherapy** uses radioactive substances placed inside or next to the tumour, away from healthy **tissue**. For example, a small pellet can be placed inside the prostate **gland**. This protects the nearby rectum from the radiation damage that a penetrating external source would cause.

The doctor may inject a radioactive liquid. Radioactive material is taken up and concentrated in the cancer but not elsewhere in the body. For example, radioactive iodine becomes one thousand times more concentrated in the thyroid gland than in other organs.

Beta particle or **gamma ray** sources are used because radiation not absorbed by cancer cells passes easily out of the body. Alpha particles are not normally used because they are extremely damaging.

**FIGURE 3**: Radiotherapy. Why has the radiotherapist fixed a cage around this patient's head?

## QUESTIONS

**3** Explain why cancer cells are more badly affected by radiation than healthy cells.

**4** Describe the advantages of internal radiotherapy over external radiotherapy.

**5** Explain why a radiotherapist leaves the treatment room before radiotherapy is started.

# Ethical issues

Doctors have to weigh up the advantages of killing tumour cells against the disadvantages of radiation damage to healthy tissues and organs or causing new cancers.

A risk–benefit analysis uses guidelines to decide the nature of acceptable treatment and whether treatment should go ahead. Three main principles are usually applied.

> Justification – is the risk justified by the benefits?

> ALARA – doses should be As Low As Reasonably Achievable yet still effective.

> Upper limits to doses – risk is unacceptable above a set limit.

## QUESTIONS

**6** What ethical problems might arise if a doctor wanted to give radiotherapy to a tumour near to a patient's ovary?

Q ... radiotherapy

# Radiation and medical imaging

**THE SCIENCE IN CONTEXT** Being able to 'see' inside the human body without the need for surgery was in the top ten medical advances voted for by medical experts. The discovery of X-rays by Wilhelm Roentgen in 1895 led to the use of X-rays, and later gamma rays, for medical imaging. People who take the images are radiographers.

## CT scans

CT (computerised tomography) uses X-rays in slices across the body to give 3D rotatable images, including soft tissues that cannot be seen using conventional X-ray imaging. Surgeons are able to see inside the body to detect and treat problems such as tumours.

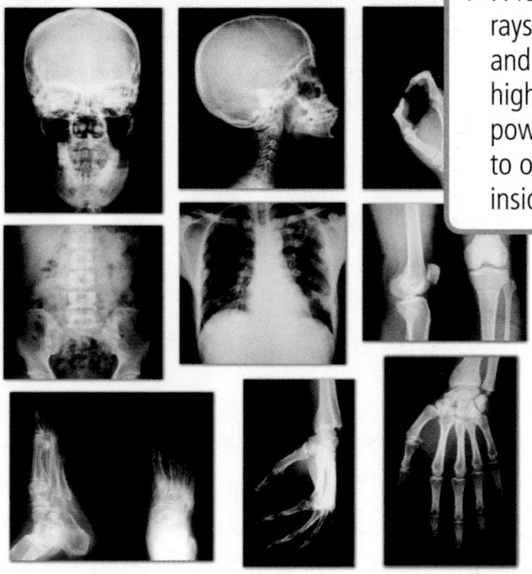

**FIGURE 1**: X-rays of various parts of the human body.

## Diagnostics

### Diagnostic radiographers and radiologists

> Diagnostic radiographers produce images for the diagnosis of injury or disease. They use X-rays and gamma rays.

> Radiologists are doctors who diagnose disease. They analyse X-rays and other scans.

### X-rays and gamma rays

X-rays and gamma rays are electromagnetic waves. They are high-frequency, high-energy, transverse waves. They can penetrate materials and ionise atoms. This makes it possible to form images of structures inside the body.

It also makes them potentially very dangerous. Why are they used in medicine if there is a safety risk? It is much better to use non-invasive techniques than to use surgery, which carries a risk of serious infection.

> X-rays can be used to produce images of broken bones, decaying teeth or diseased lungs.

> Gamma rays can be used to monitor cell activity.

### Did you know?

Experts are worried about full body X-ray scanning machines at airports that reveal hidden weapons and passengers' naked bodies. It is not for dignity's sake, but because of the increased cancer risk.

### QUESTIONS

1 (a) What type of radiation are X-rays and gamma rays?
(b) What makes them potentially dangerous?
(c) What properties make them useful in medicine? Why?

## X-rays and gamma rays

### Using X-rays

X-rays pass easily though soft tissues such as skin and muscle. They are absorbed by denser materials such as bone.

Absorption depends on the atomic number of atoms. Calcium in bone has a high atomic number (20) compared with soft tissues, which are mainly carbon (6), oxygen (8) and hydrogen (1).

**FIGURE 2**: The radiographer sits behind a lead glass screen to protect herself.

Q ... gcse x-rays ... gcse uses of radiation

X-ray opaque tracers can be used to see obstructions in the gut or blood vessels. High **atomic number** elements, such as barium, can be swallowed in a 'barium meal' or injected into a blood vessel.

X-rays are useful for investigating:

> bone fractures

> dental decay

> lung examinations

> arterial or digestive system obstructions, using X-ray opaque dyes.

### Detection of X-rays

X-rays are detected using photographic film. When the film is developed, the darkest areas show where the most electrons have been transmitted through the tissues.

### Using gamma rays

**Radioactive tracers** (radiopharmaceuticals) are weak sources of gamma rays. The doctor injects a small quantity of a solution containing radioactive material that will accumulate in the organ under investigation. Examples include:

> barium – stomach and intestinal disorders

> technetium – blood flow

> iodine – thyroid gland.

Gamma rays are useful for:

> investigating brain activity

> tumour detection and monitoring

> location of infections.

### Detection of gamma rays

The radioactive tracer is detected by a **gamma camera**. When gamma rays strike a layer of crystal, it gives off flashes of light that are detected by light detectors.

| Using X-rays | |
|---|---|
| **Advantages** | **Disadvantages** |
| widely available | can cause tissue damage and cancers |
| relatively cheap | especially harmful to unborn children and reproductive organs |
| mobile equipment can be used | not easy to distinguish soft tissues |
| quick and easy to use | 2D images |

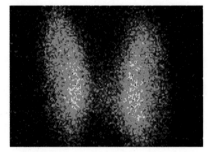

**FIGURE 3**: The thyroid glands of a patient in which radioactive iodine-123 has been injected to allow a doctor to identify a goitre (or growth).

| Using gamma rays | |
|---|---|
| **Advantages** | **Disadvantages** |
| useful for observing activity in specific tissues and tumours | expensive and gamma rays can be harmful |
| the least ionising radiation | penetrate furthest so they are the hardest to protect against |

## QUESTIONS

**2** Place these in order of darkness of image on an X-ray: fat, gas, bone, metal, muscle.

**3** Where in the body might you find (a) gas (b) liquid?

**4** Why do radiologists use lead-lined sheets?

## Choosing radiopharmaceuticals

Radioactive sources for tracers need to be selected carefully. There must be enough radioactivity at the right frequency for the camera to detect, while doing as little damage as possible. The radioisotope chosen must not stay in the body too long.

Technetium-99m has a half-life of six hours. Its effective half-life in the body, due to biological effects removing the **compound**, is about five hours.

## QUESTIONS

**5** Explain what makes technetium a good radiopharmaceutical.

# Radiation safety

**THE SCIENCE IN CONTEXT** The use of nuclear radiation, specifically gamma radiation, for radiotherapy depends on its ability to damage or destroy cells. The problem is that it cannot tell the difference between a healthy cell and a cancerous cell. So it can be very harmful as well as being very useful. People such as radiographers must protect themselves from exposure to radiation.

## Marie Curie

Marie Curie discovered two new elements, polonium and radium, and investigated the treatment of cancers using radioactive isotopes. Sadly, early investigators did not appreciate the dangers of ionising radiation, and she died from anaemia.

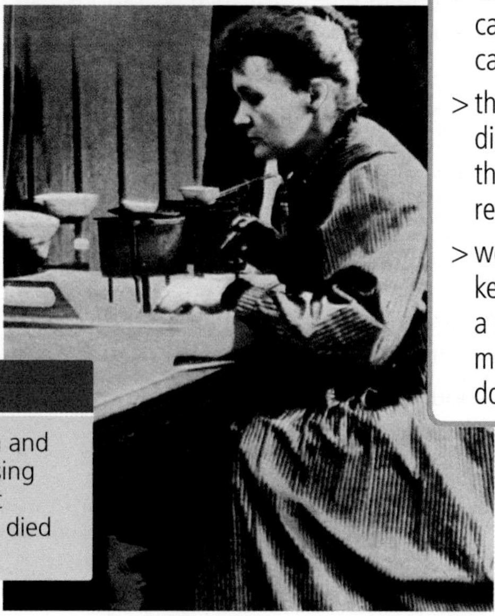

**FIGURE 1**: Marie Curie is the only person ever to win Nobel Prizes in two sciences – Physics (1903) and Chemistry (1911).

## Dangerous energy

The high energy in radiation makes it both helpful and harmful. Workers using radiation sources must make rigorous risk assessments. They follow strict health and safety guidelines.

> Radiation should be prevented from reaching the body (except for treating or diagnosing illness).

> Radioactive materials should be stored in lead-lined containers.

> Workers should wear protective gloves and clothing. They should use special shielding and work as far away from the source as possible.

> Very dangerous sources should only be handled using remote-controlled robot arms.

### Mutations

The three nuclear radiations, alpha particles, beta particles and gamma rays, can all knock electrons from atoms and ionise them. X-rays will also act as **ionising radiation**.

Inside the body, ionising radiations can alter the DNA in the nucleus of cells. A change in the genetic code is a **mutation**. This may trigger the growth of **cancer** cells.

Radiation can also kill cells. Heavy doses can damage so many cells that the body cannot replace them quickly enough. This causes **radiation sickness**. Vital organs may fail and death can occur. 70 000 people who survived the initial impact of the atomic bomb on Hiroshima died in this way.

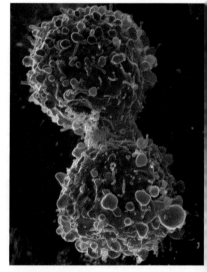

**FIGURE 2**: Radiation can ionise and damage the DNA in cells. Then, they divide uncontrollably and produce a cancerous growth.

### Did you know?

Marie Curie's notes are too dangerous to handle, they are kept in lead-lined boxes. Even her cookbook is highly radioactive.

## QUESTIONS

**1** Explain why ionising radiation is dangerous.

**2** Describe the precautions that can be taken to protect radiation workers.

Q ... gcse radiation safety

# Radiation doses

### Inside or outside?

Inside the body, perhaps after swallowing or breathing in a radioactive source, alpha radiation is the most dangerous. It is easily absorbed by cells. Beta and gamma are likely to pass straight through. Outside the body, beta and gamma are most dangerous. Unlike alpha, they can penetrate the skin and damage cells inside.

### X-rays and gamma rays

X-rays and gamma rays are used in medicine because they are high-frequency, high-energy waves with the ability to penetrate materials. This makes them potentially very dangerous.

Doses to patients should be given:

> at the lowest possible intensity

> for the minimum time possible.

Note:

> **Radiation intensity** is the energy arriving each second.

> **Radiation dose** is the amount of energy received.

### Radiation dose

The gray (Gy) is the unit of radiation dose. One gray is the transfer of one joule energy by radiation to one kilogram of body tissue.

1 Gy = 1 J per kg

### How are radiation doses monitored?

Radiographers must protect themselves from exposure to the radiation. Workers exposed to extra radiation risks have their radiation dose continuously monitored, usually with a radiation badge (see Figure 3).

A piece of photographic film is held behind a set of shields that restrict the radiation the film.

> The badge is worn for a set time.

> The film is developed.

> The amount of blackening shows the radiation absorbed by the badge and the wearer.

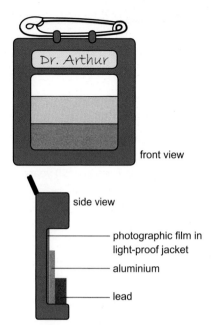

FIGURE 3: A radiation badge.

front view

side view

photographic film in light-proof jacket

aluminium

lead

## QUESTIONS

**3** Which radiation type is most dangerous inside the body?

**4** (a) Why is the film in a radiation badge partially covered by metal shields? (b) How might the radiation levels indicated be misleading?

# Safety principles

### The precautionary principle

High radiation doses can have immediate effects. Low doses increase the risk of illness in the future. As it is hard to know the effects of low doses, regulations on exposure to ionising radiation are based on the 'precautionary principle' – in the case of any doubt, act on the side of caution.

### The ALARA principle

All radiation doses should be kept As Low As Reasonably Achievable. This usually takes into account the cost involved.

Higher safety margins are more expensive. In the end our society has to decide on the acceptable level of risk.

## QUESTIONS

**5** Do the precautionary and ALARA principles make workers safe? Explain your answer.

# Preparing for assessment:
# Applying your knowledge

*To achieve a good grade in science, you not only have to know and understand scientific ideas, but you need to be able to apply them to other situations and investigations. These tasks will support you in developing these skills.*

## ✸ Greatest medical advances

What do you think has been the greatest medical advance in the past 150 years?

In 2007, the British Medical Journal launched a competition to find out. From over a hundred nominations, fifteen were shortlisted. Here are the percentages of votes each of the top ten received:

| Medical breakthrough | % of vote |
|---|---|
| sanitation | 15.8 |
| antibiotics | 14.5 |
| anaesthesia | 13.9 |
| vaccines | 11.8 |
| structure of DNA | 8.8 |
| germ theory | 7.4 |
| oral contraceptive pill | 7.4 |
| evidence-based medicine | 5.6 |
| medical imaging | 4.2 |
| computers | 3.6 |

During the 20th century, average human life expectancy increased by nearly 35 years. It has been estimated that about 30 of these were due to improved sanitation and living conditions. This is why sanitation came top of the list.

Here are just a few of the other breakthroughs:

> William Morton introduced anaesthetics in 1846

> Louis Pasteur established the germ theory in the 1860s

> Wilhelm Roentgen discovered X-rays in 1895

> Alexander Fleming discovered the antibiotic penicillin in 1928

> Crick and Watson discovered the structure of DNA in 1953.

More recently, advances include:
CT, ultrasound and MRI scans; robotic surgery; laparoscopy; endoscopy; angioplasty; heart by-pass surgery.

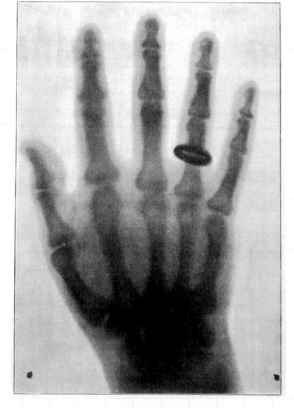

One of Roentgen's first experiments with X-rays was an image his wife's hand with a ring on her finger.

## ✹ TASK 1

(a) Sanitation (including sewage, reliably clean water, and decent housing) was top of the list. It received more votes than Nobel Prize-winning advances such as the discovery of X-rays and the structure of DNA. Suggest why this might be.

(b) A therapeutic drug is one that can be used to treat or prevent disease. Which of the top five medical breakthroughs in the list involved the development of new therapeutic drugs? Give an example of one of these.

(c) Some drugs just mask a disease – they relieve the symptoms but do not cure their cause. Give some examples.

(d) The Medicines and Healthcare Regulatory Agency (MHRA) is the government body that regulates the manufacture and commercialisation of drugs in the United Kingdom. Suggest reasons for why it takes many years after the discovery of a new drug before it can be marketed and prescribed for patients.

## ✹ TASK 2

(a) Antibiotics were second in the list. Describe the two ways in which antibiotics might work.

(b) Describe how some bacteria may become resistant to antibiotics.

(c) In recent years, the number of antibiotic-resistant bacteria has increased. Suggest a reason why you think this has happened.

## ✹ TASK 3

Vaccines can reduce the number of cases of infectious diseases. They provide you with immunity without having to first catch the disease. Explain how they do this.

## ✹ TASK 4

Wilhelm Roentgen discovered X-rays by chance. In 1895, he was studying cathode rays, a term used before scientists knew about sub-atomic particles.

(a) Name the three types of nuclear radiation, describe them and say which type cathode rays are.

(b) What type of radiation are X-rays?

(c) Roentgen's discovery was the start of medical imaging – a range of techniques that allow doctors to 'see' inside a body without the need for surgery. However, X-rays and gamma rays must be used with great caution. Explain why.

## ✹ MAXIMISE YOUR GRADE

**Answer includes showing that you can...**

**E**
- recall that disease may be treated with medicines that contain useful drugs, for example, the antibiotic penicillin.
- recall that most bacteria, but not viruses, may be killed by antibiotics.
- recall that new drugs need to undergo extensive clinical trials.

**C**
- recall that some medicines relieve the symptoms of disease, but do not provide a cure, for example, aspirin and other painkillers.
- recall that X-rays and gamma rays are examples of transverse waves and a form of electromagnetic radiation.
- recall that using of high-energy radiation can be dangerous and needs to be monitored.
- state the characteristics and properties of the three main types of nuclear radiation.

**A**
- explain how vaccination protects humans from infection.
- describe how some bacteria develop resistance to, or may not be easily treated by antibiotics.

199

# Unit 3 Theme 1 Checklist

## To achieve your forecast grade in the exam you will need to revise

Use this checklist to see what you can do now. Refer back to the relevant pages in this book if you are not sure. Look across the three columns to see how you can progress. **Bold** text means Higher tier only.

Remember that you will need to be able to use these ideas in various ways, such as:

> interpreting pictures, diagrams and graphs

> applying ideas to new situations

> explaining ethical implications

> suggesting some benefits and risks to society

> drawing conclusions from evidence that you are given.

Look at pages 250–271 for more information about exams and how you will be assessed.

| To aim for a grade E | To aim for a grade C | To aim for a grade A |
|---|---|---|
| Know that most drugs are legal and used to cure or prevent disease.<br><br>Know that some drugs are illegal and some are used for recreational purposes. | Know that drugs may cause side effects if they are overused.<br><br>Recall examples of recreational drugs that may harm the body. | Know that testing for illegal drugs is carried out in some workplaces to improve the health and safety of employees. |
| Recall that disease may be treated with medicines that contain useful drugs. | Explain that some medicines, including painkillers, help to relieve the symptoms of disease, but do not provide a cure. | Understand the personalisation of medicines. |
| Recall that some people may become dependent on or addicted to recreational drugs. | Describe how tobacco smoke causes diseases of the respiratory and circulatory systems.<br><br>Describe how alcohol affects the nervous system. | Describe evidence that links respiratory and circulatory disorders to the misuse of tobacco and alcohol. |
| Recall that most bacteria, but not viruses, may be killed by antibiotics. | Know that pathogens mutate spontaneously, producing resistant strains.<br><br>Describe how some bacteria develop resistance to, or may not be easily treated by, antibiotics.<br><br>Describe the problems caused by over-prescribing of antibiotics. | **Explain how resistant strains develop: antibiotics kill individual pathogens of the non-resistant strain, individual resistant pathogens survive and reproduce, so the population of the resistant strain rises.**<br><br>**Know that antibiotics are not used to treat non-serious infections such as mild throat infections in order to slow down the rate of development of resistant strains.** |

| To aim for a grade E | To aim for a grade C | To aim for a grade A |
|---|---|---|
| Know that, before a medicine can be used for treating a disease, it undergoes extensive trials. | Know that, before new medicines can be released onto the market, extensive research is carried out in laboratories and in clinical trials. Recall that new medicines must be passed by the Medicines and Healthcare Regulatory Agency (MHRA). | Understand issues concerning the testing new drugs on animals and humans. |
| Know the names of some diseases caused by bacteria. | Describe how pathogens can enter the body. Describe how platelets help to form a barrier to infection through a cut. Describe how bacteria and viruses can make us feel ill. | Describe how phagocytes and lymphocytes help to defend against pathogens. |
| Describe how antibodies in the blood provide immunity to certain diseases. | Explain what is meant by immunity. Explain how vaccination protects humans from infection. | Discuss the value to individuals and populations of being vaccinated against diseases. Describe how the occurrence of diseases has changed as a result of use of vaccinations. |
| Recall that X-rays and gamma rays are a form of electromagnetic radiation. | Recall that X-rays and gamma rays are examples of transverse waves. | Understand that ionising radiation kills living cells and therefore can be used to treat cancer. |
| Recall the characteristics and properties of the three main types of nuclear radiation. | Describe the characteristic properties of X-rays that allow them to be used to diagnose medical disorders. Know that some medical imaging equipment uses gamma rays, which can be detected using a gamma camera. | Describe how both external and internal radiation may be used for diagnosis. Discuss the advantages and disadvantages of using ionising radiation for the diagnosis and treatment of diseases. |
| Know that the use of high-energy radiation can be dangerous and needs to be monitored. | Explain why people who work with radiation wear film badges. Describe the construction of a film badge. | Describe the penetrating power and hazards of alpha, beta and gamma radiation. |

## What you should know

### Electrical circuits

Electric current is a flow of charge.

Current is not used up as it flows around a circuit.

 What does a battery do in a circuit?

### Energy

Energy does not disappear when it is used.

Energy can be transferred.

 Describe three ways in which energy is transferred.

### Genetics

Genes control characteristics.

Genes are made out of DNA.

 Where is your DNA found?

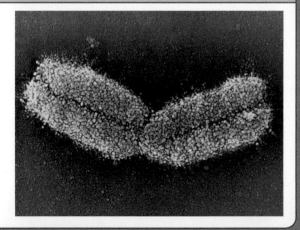

# You will find out

## Uses of electroplating

> Electroplating coats conducting surfaces with a protective or decorative layer of metal.

> Electroplating is carried out in an electrolysis cell.

> In electrolysis, reactions take place at the cathode and at the anode.

## Developing new products

> Smart paint can heal its own scratches.

> Superconductors have almost zero resistance at low temperatures.

> Thermochromic materials change colour in response to changes in temperature.

> Photochromic materials change colour in response to changes in light intensity.

## Selective breeding and genetic engineering

> Selective breeding can be used to increase food supplies.

> Selective breeding can reduce variety and cause unfavourable characteristics to appear.

> Genetic engineering can be used to increase food supplies and treat diseases, but has potential risks.

# Electroplating

**THE SCIENCE IN CONTEXT** In the presence of water and air, iron rusts and many other metals also corrode. However, these metals can be protected by electroplating. Electroplating is also used to decorate objects. There are many uses of electroplating. Unfortunately, the process involves electricity and hazardous chemicals: the electroplating industry works to minimise the risks from these hazards.

### Essential notes

> electroplating coats metallic or other conducting surfaces with a layer of metal by electrolysis
> electroplating protects and decorates surfaces
> electroplating is carried out in an electrolysis cell
> electroplating involves the movement of charged particles in an electrolyte

### Jewellery makeovers

One use of electroplating is for jewellery makeovers. Popular choices are to coat stainless steel with gold, gold with platinum, or copper with silver.

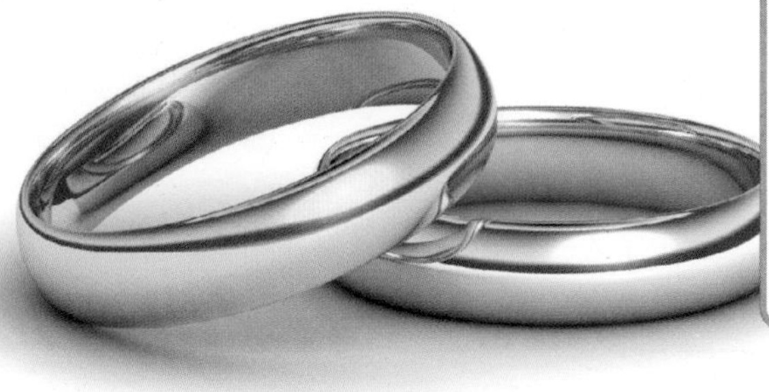

**FIGURE 1**: A stainless steel ring can be electroplated wih gold.

## Protect and decorate

Many metals corrode in the presence of air and water. Electroplating can provide protection against **corrosion**.

**Electroplating** is the process of coating a metallic or other conducting surface with a thin layer of metal by **electrolysis**. Occasionally, the electroplated layer is an **alloy** such as **brass** or **bronze**.

Many everyday items made from iron are electroplated with zinc (this is called galvanisation). Other items, such as food containers, may be electroplated with tin.

Chrome plating is common. Lots of items are coated with a layer of chromium (which is where the word 'chrome' comes from). Chromium not only protects against corrosion, but is also **decorative** – many people like its bright shiny silvery appearance. Car and motorbike parts and many household fittings are examples.

Cutlery may be coated with silver by electroplating – for protection and for appearance. In the food industry, cooking utensils are often electroplated with tin.

Some things such as jewellery are electroplated purely for decoration. Gold and silver is the usual choice. In the case of nickel jewellery, coating it with precious metals can help protect sensitive skin against allergic reactions (Figure 3).

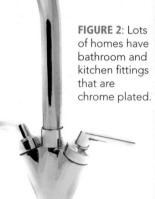

**FIGURE 2**: Lots of homes have bathroom and kitchen fittings that are chrome plated.

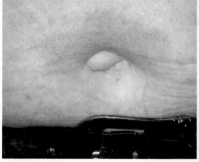

**FIGURE 3**: This rash was caused by an allergic reaction to the nickel buckle.

### QUESTIONS

1 Give two reasons why electroplating is used.

2 What metal is deposited in chrome plating?

3 Name a metal that is used for electroplating equipment used in the food industry.



Q ... chemistry of electroplating ... uses of electroplating

The reactions that h
ionic equations.

At the cathode, met
number of electron:
For example:

$Ag^+ + e^- \rightarrow Ag$

$Cu^{2+} + 2e^- \rightarrow Cu$

A general equation

At the anode, meta
through the circuit)
through the electro
the cathode. For exa

$Ag \rightarrow Ag^+ + e^-$

$Cu \rightarrow Cu^{2+} + 2e^-$

The general equatic

### ● QUESTIO

**4** Write an equa
zinc ions, $Zn^{2+}$.

**5** Describe how
electroplating.

## Reactivity

When solutions of r

> the metal deposit

> hydrogen gas is p

> both the metal ar

It depends on wher
hydrogen in the ser
the metal is deposit
**attraction** of metal

For metals just belo
produced – though
more concentrated

For two metals belc
far apart in the rea
deposits first. Agair

### ● QUESTIO

**6** Imagine you l
Think about the

# Electrolysis and the electroplating process

Suppose you have a ring or metal chain made of steel.
You want to coat it with silver to make an attractive
and expensive looking piece of jewellery.

To electroplate it, you would need:

> a strip of the plating metal (such as silver)

> solution containing ions of the plating metal (such as
silver nitrate solution)

> d.c. electricity supply, such as a battery.

In the electric circuit shown in Figure 4:

> the steel chain is the **cathode** – the negative
**electrode**

> the silver strip is the **anode** – the positive electrode

> the solution of silver nitrate is the **electrolyte** –
a liquid that can conduct electricity.

Positively charged silver ions move through the
solution to the metal chain. At the chain, they gain
electrons to form silver atoms. Silver deposits on
the chain. The silver strip dissolves to replace the
silver ions.

## Industrial processes

The electroplating industry takes great care to manage
risks. The chemicals used include some that are
corrosive and toxic. There is a lot of 'wet chemistry'
with the potential for spillage and leakage.
Low-**voltage**, high-**current** d.c. electricity is used.

The Health and Safety Executive (HSE) provides
guidelines on workplace welfare. It warns that "the
combination of electricity, water, damp and corrosive
conditions can be lethal".

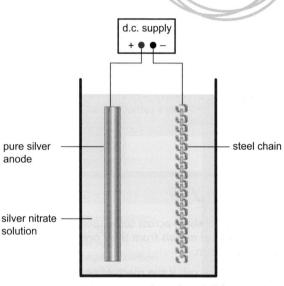

**FIGURE 4:** An electrolysis cell for
plating a steel chain with silver.

### ● QUESTIONS

**4** In electroplating, is the object to be plated
the anode or the cathode?

**5** What reaction happens at the cathode?

**6** Explain why silver ions flow in the opposite
direction to electrons (Figure 4).

## Plating plastics

Most **plastics** do not conduct electricity. However, there are ways to
electroplate them. Here is an example.

In the car industry, plastic parts are cleaned to remove any oil and grease.
They are then treated with a solution that etches their surfaces, making
tiny pits all over it. It is rather like using rough sandpaper to prepare
paintwork for repainting.

The etched plastic parts are washed and put into palladium chloride
solution. Palladium metal deposits in the tiny pits. This makes the surface
conducting and copper can be electroplated onto it. In turn, this copper
can be plated with chromium or other metals such as nickel or gold.

### ● QUESTIONS

**7** Explain why plastics
cannot be electroplated
directly.

**THE SCIENC**
working in th
apply knowle
and electrolys
applications b
used, for exar
purification of

**Cleaning**

Mining for or
metal pollutic
Electrolysis ca
plating the m
can be recove

**Electro**

You will have
purification of
and decoratio

An electrolysis
out. It is conne

> The power s
Battery term

> The cell has
positive elec
terminal of
terminal of

In electropla
terminal. It i

> The cell con
be a **molten**
water. In bo
negative ion
through the

FIGURE 2: An
electrolysis ce

a mat

It may
or a solution of

# Preparing for assessment: Planning a practical investigation

*To achieve a good grade in science, you not only have to know and understand scientific ideas, but you need to be able to apply them to other situations and investigations.*

Connections: This task relates to Unit 3 context 3.5.2.1 Uses of electroplating. Working through this task will help you to prepare for the practical investigation that you need to carry out for your Controlled Assessment.

## ✳ Investigating the purification of copper

A group of students learned that electrolysis is used to electroplate some metals. Earlier they had found out about how metals are extracted from their ores. They read that copper is purified by electroplating it on to a piece of pure copper.

It is important that industrial processes use the most efficient conditions – those that maximise the yields as quickly as possible (as time means money). When copper is purified from impure copper, using an electrolyte of copper sulfate, time, current and the concentration of the electrolyte are all important to ensure that the best yields are achieved.

The students decided to investigate the conditions which affect the amount of copper deposited when copper(II) sulfate is electrolysed.

## ✳ Research

**1.** Find out how copper is purified by electrolysis and what conditions affect the yields.

**2.** Research the purification of copper using electrolysis.

Use at least three sources of information (make sure that you list them). Use the key words: electrolysis, purification of copper, uses of copper. Collect together all your research and reference it correctly.

You will need to understand anode, cathode, electrolyte, current, ions. Reference everything that you find out.

The students were shown how to find the mass of copper deposited during electrolysis using the apparatus shown. Using this method, they carried out a number of trial experiments.

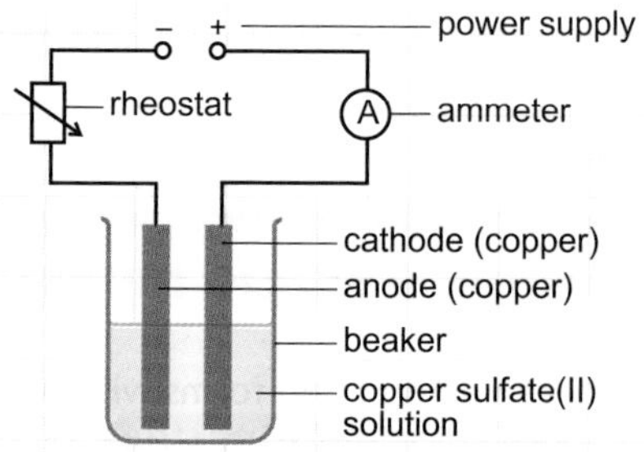

> They weighed the copper cathode, recorded its mass and connected it to the circuit.

> They adjusted the rheostat so that a current of 0.5 A passed and started a stopclock.

> They turned off the power supply after 30 minutes, removed the cathode and washed it carefully with running water and then a little propanone.

> When it was dry, they re-weighed the cathode to find the mass of copper deposited.

While carrying out these experiments the students noticed:

> the rheostat needed to be adjusted regularly to keep the current constant

> the copper deposited on the cathode fell off easily.

power supply
rheostat
ammeter
cathode (copper)
anode (copper)
beaker
copper sulfate(II) solution

They were also warned that, if the cathode was not dried properly after electrolysis, small amounts of water would add to the mass of the cathode.

The students decided to adapt this method to investigate factors that might affect the quantity of copper obtained. Importantly, they wanted to find conditions that improved how well copper sticks to the cathode.

Some data obtained by the students in these experiments:

| Experiment | Mass of copper deposited on the cathode (g) | Current (A) | Time (minutes) |
|---|---|---|---|
| 1 | 0.16 | 0.25 | 30 |
| 2 | 0.30 | 0.25 | 60 |
| 3 | 0.30 | 0.5 | 30 |
| 4 | 0.59 | 0.5 | 60 |

## ✳ Planning

**3.** Imagine you are one of the group and are leading the other students. Suggest a hypothesis that the investigation will test.

Give the purpose in terms of finding out about the relationship between two variables. You might investigate more than one pair of variables.

**4.** Make notes about how you would set about the investigation using your research to help you. Also look carefully and use the outcomes of the earlier experiments carried out by students to help your planning.

Include a summary of the experiments that the groups will carry out. Include a list of equipment and materials needed with the sizes of the equipment and how much of the chemicals. Identify independent, dependent and control variables. Say whether the variables are continuous or categoric. For each pair investigated, identify independent and dependent variables. Also, identify the control variables (the other variables that will be kept constant).

**5.** Write a clearly structured step-by-step plan for the investigation.

**6.** For any pair of variables (independent and dependent), explain carefully the relationship between two variables that you are planning to measure and how you will look for any patterns in this relationship.

The plan should be written in sufficient detail and clarity so that it can be followed easily by another person or group. Include your equipment and chemical list. Remember to give instructions on the range of readings you need and the time interval between them.

How will data be recorded? What methods will you use to look for trends and patterns?

## ✳ Assessing and managing risks

**7.** Look at your plan for the investigation, list any potential hazards. Give the associated risks.

Make sure that you say what the risks are, for each hazard that you list.

**8.** Explain how you would minimise the risks.

Give the scientific reasoning for the steps you suggest to minimise risks.

# Selective breeding

## Essential notes

> selective breeding can be used to increase food supplies

> selective breeding can reduce variety and may cause unfavourable characteristics to appear

**THE SCIENCE IN CONTEXT** Feeding the world's rapidly growing population is a major challenge. Agricultural scientists have used selective breeding to produce animals and crops with particular characteristics, such as higher yields and greater resistance to disease, in order to provide more food for the planet's increasing population.

### Wheat breeds

Wheat plants have been selectively bred over thousands of years to give high-yielding, disease-resistant varieties with varying protein content. It makes the flour suitable for a wide variety of purposes, from baking bread to making pasta. The fight to resist fungus diseases goes on.

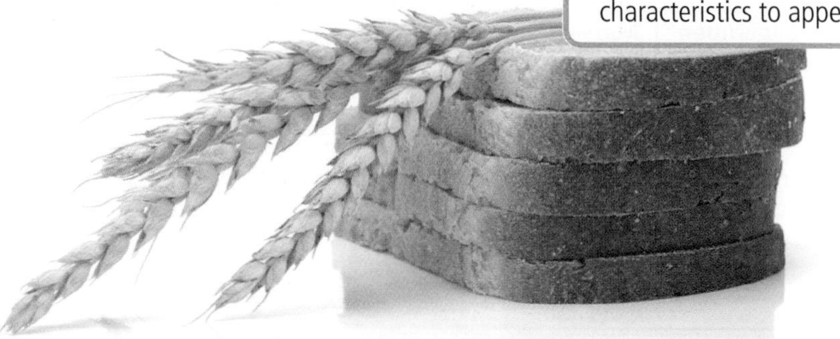

**FIGURE 1**: High-performing modern wheat varieties are the product of many centuries of selective breeding.

## Selective breeding

### World population

About one billion people have insufficient food. From 2000 to 2050, the world population will probably rise from about six billion to about nine billion. More than a 50% increase in food production will be needed.

Agricultural scientists have the daunting task of improving the world's food supply, so that the number going hungry does not rise dramatically.

### Selective breeding

The traditional method for improving crops and livestock, for characteristics such as increased size or better disease resistance, is **selective breeding**.

In ten thousand years of farming, many new varieties of animals and plants have been produced. These have hugely increased yields over the original wild species.

### How it works

Breeders improve their breeding stock by selecting only those animals and plants with the best characteristics. Selective breeding is known as artificial selection.

> Choose the desired characteristic.

> Select the best animals or plants.

> Breed only from them.

> Select offspring and breed only from the best.

> Continue the selection process over many generations.

Qualities such as disease resistance in wheat, milk yields in dairy cattle and meat yields in pigs can be improved gradually over many generations.

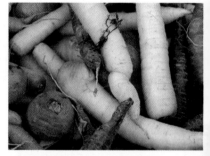

**FIGURE 2**: American scientists have selectively bred multicoloured carrots.

### Did you know?

The World POPClock projection estimated the number of people on planet Earth at 20.00 Greenwich Mean Time on 25 October 2010 to be 6 877 371 663.

### QUESTIONS

**1** Why is it necessary to improve crop plants and animals?

**2** Explain why selective breeding involves artificial selection.



... world food day ... selective breeding

# Risks of selective breeding

Most of today's healthy, disease-resistant and highly productive farm animals and plants have been produced over centuries of selective breeding. However, using the technique can cause problems.

## Inbreeding

Breeding closely related animals and plants greatly increases some risks:

> Hereditary diseases caused by recessive genes may appear. Individuals are more likely to have recessive genes that would normally be masked by dominants present in pairs.

> Whole populations may be badly affected by the same disease. Individuals that are more alike are more likely to be susceptible to the same disease.

> Many **genes** may be lost. There will be a reduction in the gene pool for the species – it will be more difficult to produce new varieties in the future.

## Development of unfavourable characteristics

Unfavourable characteristics may appear. Generations of inbreeding in pedigree dogs has led to some serious health problems. These include pugs with breathing problems and boxers with epilepsy.

**FIGURE 3**: Short-nosed pugs often have breathing problems.

### Did you know?

Inbreeding increases the risk of recessive hereditary disorders in cattle, such as BLAD, DUMPS and MSUD.

These abreviations stand for bovine leucocyte adhesion deficiency, deficiency of uridine monophosphate synthase, and maple syrup urine disease (MSUD).

### QUESTIONS

**3** Inherited disorders are more likely to appear in a population that has been selectively bred. Explain why.

**4** Which disadvantage of selective breeding do you think is the most important? Explain your answer.

# Breeding for disease resistance

Wheat rust is a persistent problem for farmers.

> In California, 25% of the wheat crop was lost to stripe rust in 2003.

> The fungus mutates and evolves rapidly to overcome new resistant strains of wheat.

Geneticists and plant breeders constantly have to use selective breeding to produce new wheat varieties and so keep ahead of the evolving fungus.

Worm parasites can badly affect farm animals such as sheep. The parasites become resistant to chemicals.

In Australia, scientists have succeeded in breeding Merino sheep for resistance to worms with little or no chemical treatment.

**FIGURE 4**: Selective breeding can increase resistance to diseases such as wheat rust parasites.

### QUESTIONS

**5** Describe the effect that selective breeding has on variety.

**6** Scientists have started to conserve animal and plant species and varieties, to maintain a global gene bank with as much variety as possible. Explain why.

# Genetic engineering

**THE SCIENCE IN CONTEXT** Biotechnologists and geneticists use genetic engineering to transfer genes between species. They use cloning techniques to produce new plants, tissues and organs rapidly. They improve the yields and nutritional value of food crops, use bacteria to make drugs and vaccines, and create animal and human tissues for replacement surgery.

## Mending muscles

A research team has created a genetic switch that allows light signals to be switched on in muscle stem cells. Using mice, they have been able to pinpoint exactly how stem cells turn into muscle cells. This could lead to a drug that enables people to grow new muscle cells to replace worn out or damaged cells – for example in muscular dystrophy.

**FIGURE 1**: Genetically engineered mice are being used to investigate the details of how mammalian bodies work.

## How to engineer a gene

DNA molecules are always made out of the same basic units. You can take a gene which is foreign to one species and transfer it to another. This is genetic engineering.

### Genetically engineered insulin

It is possible to transfer the gene for making human insulin from a human cell into a bacterial cell.

The bacterium is now **genetically modified**. It is a GM bacterium. The bacterium will make human insulin.

When the bacterium divides, the DNA will be copied. The bacteria are cloned – they all have identical genetic makeup. All the offspring will be able to make human insulin.

People who suffer from **diabetes** do not make enough of their own insulin. They inject insulin to keep their blood sugar levels down. If you grow GM bacteria in large tanks, called bioreactors, the insulin can be extracted for diabetes sufferers to use.

Human growth hormone and vaccines are also made this way.

**Did you know?**

90% of soya beans and 63% of the sweetcorn grown in the USA are genetically modified.

**QUESTIONS**

**1** Describe what a genetically modified bacterium is.

**2** Use Figure 2 to explain how it is possible for bacteria to make human insulin.

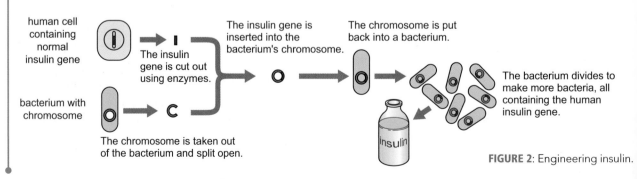

human cell containing normal insulin gene

The insulin gene is cut out using enzymes.

The insulin gene is inserted into the bacterium's chromosome.

The chromosome is put back into a bacterium.

bacterium with chromosome

The chromosome is taken out of the bacterium and split open.

insulin

The bacterium divides to make more bacteria, all containing the human insulin gene.

**FIGURE 2**: Engineering insulin.

# GM crops and cloning

## Genetically modified (GM) crops

Weeds reduce crop yields by competing for light, water and nutrients. Glyphosate is a very effective weedkiller, widely used by farmers, but it kills normal crop plants, too.

GM soya bean plants have been genetically engineered to have a glyphosate-resistant gene. Farmers can spray fields with glyphosate weedkiller without harming the soya bean crop.

To make glyphosate-resistant soya beans:

1. Choose a plant that is resistant to glyphosate weedkiller.

2. Use enzymes to cut out the gene for resistance.

3. Transfer the gene to soya **embryo** cells and grow them into adult plants.

4. Spray with glyphosate weedkiller.

5. Harvest beans from the surviving plants.

6. Plant the beans and grow glyphosate-resistant plants.

7. Increase yields by killing the weeds, but not the crop, with glyphosate weedkiller.

Other GM crops include golden rice with added vitamin A.

Using the Bt-toxin gene stops insects from eating crops such as peanut plants. Reducing insecticide use makes food cheaper to grow.

**FIGURE 3**: GM soya beans are used as animal feed and added to many food products.

## Cloning

When a useful GM organism has been made, it is important to get as many identical offspring as possible, as quickly as possible.

Biotechnologists use plant tissue **culture** (micropropagation) to produce, rapidly, a **clone** of genetically identical plants. These can be grown, for food, for example.

The same can be done with animal or human tissues. Fragments of tissue from a donor parent plant or animal can be transferred to a culture plate. The plate supplies food and oxygen for the cells to grow. It is possible to grow skin for skin grafts this way, or cartilage to repair noses or ears.

### QUESTIONS

**3** Explain why resistance to weedkillers is a useful characteristic for a crop plant.

**4** Explain why cloning, using tissue culture, is useful to the developers of GM organisms and tissues.

# Frankenfoods?

People (including some scientists) are worried about scientists 'meddling with nature'. There has been resistance to GM foods because of fears that there may be unknown side effects from eating them. So far, there is no evidence of this.

Of greater concern is the possibility that weedkiller genes could be transferred to weeds, making weedkillers useless. Perhaps the insect-resistant Bt-toxin gene could reach wild plants, so that insect food supplies and natural communities would be disrupted.

## Gene therapy

Researchers are also trying to replace harmful genes with healthy ones, for example in sufferers from cystic fibrosis.

Exchanging human genes opens up the possibility of 'designer babies'.

### QUESTIONS

**5** Would you eat GM food? Explain your answer.

**6** Discuss the issues raised if parents could 'design' their babies.

Q ... tissue culture ... gm food

# Unit 3 Theme 2 Checklist

## To achieve your forecast grade in the exam you will need to revise

Use this checklist to see what you can do now. Refer back to the relevant pages in this book if you are not sure. Look across the three columns to see how you can progress. **Bold** text means Higher tier only.

Remember that you will need to be able to use these ideas in various ways, such as:

> interpreting pictures, diagrams and graphs

> applying ideas to new situations

> explaining ethical implications

> suggesting some benefits and risks to society

> drawing conclusions from evidence that you are given.

Look at pages 250–271 for more information about exams and how you will be assessed.

| To aim for a grade E | To aim for a grade C | To aim for a grade A |
|---|---|---|
| Recall some household objects that are electroplated to prevent corrosion. | Know reasons for electroplating metals.<br><br>Explain why nickel jewellery is electroplated with precious metals. | Understand the potential risks to employees in the electroplating industry. |
| Recall that electrolysis involves the movement of charged particles in an electrolyte. | Know that the cathode is the negative electrode and the anode is the positive electrode in an electrolysis cell.<br><br>Understand that charged particles are called ions and that ions are atoms which have either lost or gained an electron. | |
| Recall what is meant by 'electroplating'. | Understand that an article to be electroplated is made the cathode, and immersed in an aqueous solution containing ions of the required metal.<br><br>Know that the anode is usually a bar of the metal used for plating. | Understand the suitability of different metals for electroplating items.<br><br>**Know how to complete simple equations to show the process at the cathode and anode.** |

## To aim for a grade E   To aim for a grade C   To aim for a grade A

| | | |
|---|---|---|
| Recall examples of products made from 'new' materials. | Describe uses for smart paints, superconductors, smart materials and chromic materials. | Explain how the properties of new materials make them better suited to their job than traditional materials in both the home and the wider world. |

| | | |
|---|---|---|
| Know that breeding programmes are used to improve the characteristics of animals and plants. | Describe how selective breeding of plants and animals is carried out.<br><br>Explain some of the advantages and disadvantages of selective breeding. | Understand the ethics of genetic engineering compared to selective breeding. |

| | |
|---|---|
| Recall that clones are genetically identical individuals. | Understand that cloning techniques involve laboratory processes to produce offspring that are genetically identical to the donor parent.<br><br>Describe tissue culture (micropropagation).<br><br>Describe uses of tissue culture including cloning plants for food and the culture of animal and human organs. |

| | |
|---|---|
| Recall that genetic engineering involves the transfer of genes. | Explain how genetic engineering involves the transfer of 'foreign' genes into the cells of animals or plants at an early stage in their development so that they develop desired characteristics. |

| | | |
|---|---|---|
| Know that genetic engineering can be used to make medicines and treat inherited conditions. | Describe how human insulin is produced using genetically modified bacteria. | Understand some of the economic, social and ethical issues concerning genetic engineering, genetically modified foods and 'designer babies'.<br><br>Understand the ethics of gene replacement therapy. |

## What you should know

### Pollution

Pollutants are harmful to living organisms.

Carbon dioxide is produced when fossil fuels are burned.

Carbon dioxide traps heat that has reached Earth from the Sun.

An increase in carbon dioxide in the atmosphere may lead to global warming.

Pollution levels in water and air can be monitored using indicator species.

 Why is the amount of carbon dioxide in the atmosphere increasing?

### Energy transfer

Energy does not disappear when it is used.

Energy may be transferred by conduction, convection and radiation.

 Describe examples of energy transfer by each of conduction, convection and radiation.

# You will find out

## Environmental concerns when making and using products

> Making and using products may increase the amount of greenhouse gases in the atmosphere.

> Greenhouse gases absorb long-wave radiation from Earth's surface, causing the atmosphere to warm up.

> Fertilisers, pesticides and herbicides can cause water pollution.

> Invertebrates can be used to monitor water pollution and lichens can be used to monitor air pollution.

> Certain plastics are degradable.

## Saving energy in the home

> Insulation can be used to minimise heat loss.

> U-values measure the rate of heat loss through a material.

> Payback time is how long it takes to repay set-up costs from energy savings made.

## Controlling pollution in the home

> Pollution-free air in the home is important for health and wellbeing.

> Indoor pollution can be more serious than outdoor pollution.

> Use of household products has associated hazards and risks.

> Risks may be minimised by adopting safe procedures or using less toxic products.

> Domestic boilers need an adequate air supply to work efficiently.

> Incomplete combustion of fuels causes lower energy output and formation of toxic products.

> Radon may become a pollutant if rocks below a home contain radium or uranium in large enough concentrations.

# Global warming

**THE SCIENCE IN CONTEXT** All of us depend on natural and manufactured products to sustain and improve our lives. Quality and cost of products is important, but many people also worry about the effect that production and disposal may have on the environment. People use scientific understanding to make products more environmentally friendly.

## Global dimming

Tiny particles of soot, ash and other pollutants in the air cause less sunlight to reach Earth's surface. Evidence for this was reported first in 2001. Later work confirmed it. There are concerns that it may have led scientists to underestimate the true significance of greenhouse gases.

### Essential notes

> making and using products may increase the amount of greenhouse gases in the atmosphere and this may link to global warming

> greenhouse gases absorb long-wave radiation from Earth's surface, causing the atmosphere to warm up

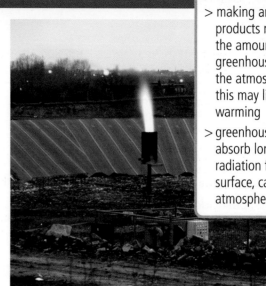

FIGURE 1: Methane is produced in landfill sites from rotting rubbish.

## Greenhouse gases

**Carbon dioxide** in the atmosphere keeps Earth warm enough for life to survive. The Sun warms Earth's surface, which then radiates energy that carbon dioxide traps. Other gases can also trap radiated energy. They are called **greenhouse gases**. **Methane** and nitrous oxide are examples.

Greenhouse gases get into the atmosphere as a result of natural processes and human activity.

### Natural processes

**Natural processes** include **respiration** and **decay**. Respiration produces carbon dioxide:

glucose + oxygen → carbon dioxide + water

Decay is a complex process. Bacteria break down dead plants and animals. Methane and carbon dioxide may be produced – how much of each depends on the conditions.

### Human activity

**Human activity** includes burning fossil fuels, agriculture and landfill sites for waste disposal.

> Carbon dioxide is produced from burning fossil fuels in power stations and vehicles.

> Methane is produced by livestock such as cows, and in landfill sites.

> Nitrous oxide comes from vehicle exhausts and power stations and increased use of nitrogen-based fertilisers.

It is important to keep Earth's surface warm enough to sustain life. However, increasing greenhouse gases in the atmosphere causes **global warming**.

Because of concern about global warming, nations have tried to reach agreements about reducing emissions of carbon dioxide, methane and nitrous oxide.

FIGURE 2: Bacteria in cows' stomachs produce methane as food is digested. Estimates of how much methane a cow burps each day range from 100 dm³ to 500 dm³.

## QUESTIONS

**1** Name two natural process that produce greenhouse gases.

**2** Describe how nitrous oxide gets into the air.

Q ... gcse greenhouse gases

# Gases and absorption of radiation

## Methane, CH$_4$

In water, bogs and swamps – where there is little air – plants and animals decay as bacteria break down organic compounds and produce methane.

The bacteria are also found in the digestive systems of some farm animals.

## Nitrous oxide, N$_2$O

The need to produce more food for the world's population has led to more intensive farming and, as a result, increased use of nitrogen-based fertilisers. There are two processes:

> **Nitrification** – bacteria in the soil convert nitrogen-based fertilisers to nitrates.

> **Denitrification** – other bacteria convert nitrates to nitrogen, nitrogen dioxide and nitrous oxide.

## Long-wave radiation

Radiation from the Sun is absorbed by Earth's surface and re-emitted with longer **wavelength** (lower energy). This long-wave infrared radiation:

> passes through the **atmosphere** and escapes back into space, or

> is absorbed by greenhouse gases.

The more long-wave radiation absorbed by greenhouse gases, the warmer the atmosphere becomes.

FIGURE 3: Rice is the major food source for 40% of the world's population. Rice fields are a perfect environment for bacteria to produce methane.

FIGURE 4: It is not just synthetic fertilisers that cause problems. Animal waste fertiliser can lead to nitrous oxide being emitted from agricultural land.

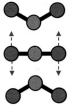

linear molecule wth average position of carbon midway between two oxygens

FIGURE 5: Carbon dioxide absorbs infrared radiation because of the way its molecules can vibrate. One way is shown above – there are others.

## QUESTIONS

**3** Describe the role of bacteria in global warming.

**4** Explain why growing rice contributes to global warming.

**5** Make models of carbon dioxide to demonstrate the vibrations shown in Figure 5. Suggest reasons why nitrogen and oxygen are not greenhouse gases.

# International agreements

For at least twenty years, nations have recognised the need for action to reduce greenhouse gases.

The Kyoto agreement came into force in 2005. It was ratified by 183 countries and the European Community. The USA refused to sign. Australia did not sign until 2007. It was agreed that industrialised countries should cut greenhouse gas emissions by an average of 5.2% from the 1990 level, by 2012.

The aim of the Copenhagen **Climate Change** Summit, held in December 2009, was to agree something to replace the Kyoto agreement. It did not produce the historic deal that many were hoping for. However, the Kyoto agreement was preserved and the debate about tackling greenhouse gas emissions is still alive.

## QUESTIONS

**6** Find out more about the Kyoto agreement and how cuts for individual countries were decided.

# Water and air pollution

**Essential notes**

> fertilisers, pesticides and herbicides can leach into lakes and rivers causing pollution, including eutrophication

> pollution levels in water and air can be monitored using indicator species

**THE SCIENCE IN CONTEXT** It is almost impossible to make, use and dispose of products without affecting the environment. It's true of manufacturing and agricultural industries. Much is being done to improve processes and products to minimise their environmental impact. Environmental scientists monitor air and water quality – looking for evidence of pollution, its source and how it can be cleaned up.

## Invisible pollution

In March 2010 a government committee published a report suggesting air pollution on UK streets contributes to tens of thousands of early deaths each year – more than passive smoking, traffic accidents or obesity. The likely cause is pollutants from cars and lorries.

**FIGURE 1:** Invisible pollutants escape into the air.

## Transport and industrial activity

Human activity has led to pollution of water and air.

Making products and burning fossil fuels both affect the environment.

Pollution can happen in rural areas as well as urban and industrial areas.

### Fertilisers and pesticides

Providing sufficient food for the world is a major challenge for the agricultural industry. Increasing crop yields is vital. Two ways to do this are:

> feeding plants with **fertilisers** (these may be **synthetic** or **natural**)

> protecting plants with **pesticides** (for example, insecticides, herbicides and fungicides).

However, care is necessary when using fertilisers and pesticides. They can dissolve in rainwater and be washed into lakes, streams and rivers. This is **leaching**. Why is it a problem?

Fertilisers provide nutrients for plants. Fertilisers that leach into lakes and rivers increases the concentration of nutrients in the water. This is **eutrophication**. It causes algae to grow. The algae starve organisms of oxygen and sunlight.

Pesticides pollute water and can kill fish and other organisms that live in water.

**FIGURE 2:** Crops are often sprayed with pesticides to control pests and disease.

### Monitoring pollution

Environmental scientists monitor pollution.

They monitor the abundance and health of certain plants and animals. They are called **indicator species**.

Water pollution is indicated if species such as bloodworm, sludge worm, rat-tailed maggot and water louse are the only ones found.

Air pollution is indicated by the variety of types of **lichen** found. In polluted air only a few types of lichen are found.

**FIGURE 3:** The absence of shrubby, hairy and leafy lichens indicates air pollution. Crusty lichens can survive in moderately polluted areas.

## QUESTIONS

1 Name two ways that farmers can increase crop yields.

2 Explain how fertilisers are leached from soil.

3 Name an indicator species that is used to monitor air pollution.

## Eutrophication

Fertilisers contain nitrogen compounds, such as nitrates, and phosphorus compounds, such as phosphates. These nutrients help plants to grow.

One of the problems with fertilisers is that they help all plants to grow.

> When leached from the soil they seep into lakes and rivers. They cause algae to grow rapidly.

> The water becomes cloudy. Sometimes a green bloom forms on the surface of the water.

> Some plants die because of lack of sunlight. They sink to the bottom and decay. Algae die and decay.

> Bacteria feeding on the dead plants and algae use up the oxygen dissolved in the water. Fish die.

Synthetic fertilisers are not the only cause of **eutrophication**. Animal manure and sewage also cause eutrophication.

FIGURE 4: Environmental scientists use test kits to measure the concentrations of nitrogen and phosphorus compounds in water.

## QUESTIONS

4 Describe how a green bloom forms on water that has been polluted by leached fertiliser.

5 Explain why fish die as a result of eutrophication.

## Indicator species

The variety of species found in a lake or river indicates the level of pollution.

Eutrophication decreases the amount of dissolved oxygen in lakes and rivers. Only organisms adapted to these conditions will be found. They include bloodworms, sludge worms and rat-tailed maggots. If there are few species other than these, the water is likely to be polluted.

FIGURE 5: These environmental scientists are taking a kick sample to monitor the variety of organisms in the river.

| Level of dissolved oxygen | Indicator species present |
|---|---|
| low | sludge worms rat-tailed maggots bloodworms |
| medium | water louse freshwater shrimp |
| high | mayfly nymphs stonefly nymphs |

TABLE 1: Indicator species and levels of dissolved oxygen in water.

## QUESTIONS

6 Find out how environmental scientists sample lakes and rivers and measure the numbers of indicator species.

# Degrading plastics

## THE SCIENCE IN CONTEXT

Most plastics dumped in landfill sites do not degrade (decompose) – they will still be there in 100 years time. However, polymer scientists and plant biologists are trying to solve the problem by making biodegradable plastics, particularly for packaging. Understanding chemical reactions is the key to success.

FIGURE 1: Some plastic waste washes up onto beaches.

### Essential notes

> some plastics are photodegradable (by exposure to sunlight) and some are oxo-degradable (by microbes)

> water-soluble plastics can be used for shopping bags, packaging and stitches for wounds

### Floating plastic

Hundreds of millions of tonnes of plastic waste float on the oceans. Waste often accumulates in huge patches – the biggest covers an area twice the size of France. It is in the North Pacific. By studying microorgansims in the sea that can degrade plastic, scientists hope to find ways of speeding up the degradation.

## Plastic waste

Worldwide production of plastics increased from five million tonnes in 1950 to 245 million tones in 2006. What happens to plastic waste?

Options include:

> dumping it into landfill sites. Most plastics do not decay – they stay around for years.

> recycling plastics, but this is not simple – there are many types of plastic and they are difficult to separate. Mixed plastic waste is much less useful.

> burning it. This can cause air pollution and is a waste of useful plastics.

### Breaking down plastics

Much plastic does not break down. It is buried in landfill sites and stays there for years. However, some plastics do break down. This process of breaking down is called degradation. The plastic degrades into smaller and smaller fragments.

Photodegradable plastics are broken down into microscopic pieces by sunlight. They do not degrade when buried in a landfill site since sunlight cannot reach them.

Oxo-degradable plastics are biodegradable. They are broken down by microbes (bacteria) into substances that may be useful – for example as **compost**.

FIGURE 2: Jackets, hats, vests and socks can be made from recycled plastic fizzy drink bottles.

### Did you know?

Built from 12 500 empty plastics bottles, the yacht *Plastiki* sailed 15 000 km across the Pacific Ocean in 2010. The purpose was to raise awareness of the enormous amount of waste plastics floating around the oceans.

### QUESTIONS

**1** Describe three ways to dispose of plastic waste.

**2** Explain what 'degradation' of a plastic means.

Q ... plastic waste ... gcse recycling

# Degradable plastics (Higher tier)

## Photodegradable plastics

These plastics are usually made from chemicals obtained from **crude oil**.

Bonds in the polymer chain can be broken by prolonged exposure to sunlight. It is often helped by a chemical additive, which catalyses the reaction.

Once photodegradation has broken the plastic into sufficiently small fragments, the breakdown is completed by bacteria and other microbes.

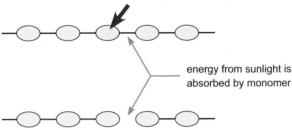

energy from sunlight is absorbed by monomer

energy from sunlight is absorbed by monomer

FIGURE 3: Monomer units in the polymer absorb energy from sunlight. It weakens and sometimes breaks bonds between monomer units – degrading the plastic.

## Biodegradable plastics

There are two main types of biodegradable plastics.

**Oxo-degradable** plastics need oxygen (from air) to degrade (by **oxidation**). An additive in the plastic ensures that the plastic breaks into fragments small enough for microbes to complete the degradation.

**Hydro-degradable** plastics are made from polymers such as polyethene and starch. In moist conditions, microbes in the soil feed on the starch between the polymer molecules. This causes the plastic to fall apart. However, fragments of plastic remain and may be not degraded further by microbes. Because of this, some people argue that hydro-degradable are not truly biodegradable.

## QUESTIONS

**3** Describe how a photodegradable plastic degrades.

**4** Explain why starch is mixed with polyethene to make a hydro-degradable plastic.

# Water-soluble plastics (Higher tier)

Some plastics, such as polyvinyl alcohol (PVOH) and ethylene vinyl alcohol (EVOH), can dissolve in water. They can be used for disposable packaging and shopping bags.

Polyvinyl alcohol is a polymer made from vinyl alcohol. It has a structure similar to polyethene except that one hydrogen on every other carbon is replaced with a hydroxyl **group**, -OH. The -OH groups attract water molecules, making it water-soluble.

Ethylene vinyl alcohol is a co-polymer of ethene (old name – ethylene) and ethenol (old name – vinyl alcohol). The polymer chain is a random mixture of ethene and ethenol monomer units in the chain.

**Remember**

The correct scientific name for vinyl alcohol is ethenol – not to be confused with ethanol.

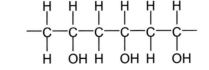

FIGURE 4 : The structure of a vinyl alcohol molecule and a section of a polyvinyl alcohol polymer chain.

## QUESTIONS

**5** Draw the structure of a section of ethylene vinyl alcohol. Suggest reasons why it might be less or more water-soluble than polyvinyl alcohol.

# Conduction, convection and radiation

**Essential notes**

> energy may be transferred by conduction, convection and radiation

> insulation can be used to minimise heat loss

**THE SCIENCE IN CONTEXT** Energy costs money, which is why architects and home owners do what they can to reduce energy losses in homes. Energy advisers and consultants need to understand how energy transfers can lead to wasted energy and what steps can be taken to minimise them. Doing so not only saves money, it also reduces the impact of buildings on global warming.

## Heatwaves

In 2010 the NHS published a Heatwave Plan for England. Among other advice, it suggests that "Council and housing associations should increase the use of reflective paint and external shading around south facing windows".

**FIGURE 1**: Buildings in very hot countries are often painted white.

## Heating a home

There are many ways to warm a room. Energy is moved from one place to another – by conduction, convection and infrared radiation.

**Conduction** is how energy is transferred through a material that is heated. Energy can also be transferred from one material to another by conduction. Some materials transfer energy more easily than others.

> **Conductors** transfer energy easily – for example, metals.

> **Insulators** do not transfer energy easily – for example, air, glass, ceramics and plastics.

**Convection** is the main way that energy is transferred in liquids and gases. Warm liquids and warm gases rise upwards because they are less dense than cooler liquids or gases. This is how hot air circulates in a room.

**Infrared radiation** is what you feel if you hold your hand next to, but not touching, a warm radiator. Unlike conduction and convection, radiation does not need a material to transfer energy through.

### Minimising heat loss

> Expanded polystyrene and mineral wool (ceramic fibres) reduce heat loss through roofs and walls and from hot water tanks.

> Double glazing works because glass and air trapped between the panes are poor conductors.

> Draught excluders prevent warm air from leaving a room.

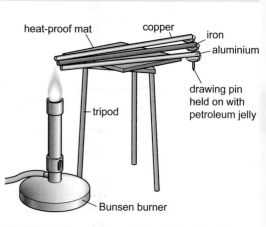

heat-proof mat — copper — iron — aluminium — drawing pin held on with petroleum jelly — tripod — Bunsen burner

**FIGURE 2**: When the petroleum jelly melts, the drawing pin drops off. How would you decide which of the three materials is the best conductor?

## QUESTIONS

**1** Describe a conductor.

**2** Explain what happens to liquids and gases when they are warmed.

**3** Describe how the transfer of energy by radiation differs from conduction and convection.

Q ... heat transfer conduction convection

# Explaining conduction and convection

In all materials, particles vibrate – they jiggle around.

> In solids, particles vibrate about fixed positions.

> In gases and liquids, particles are free to move.

## Conduction

If one end of a solid is heated, particles at that end gain energy and vibrate more. They shake their neighbouring particles. The vibrations are passed from particle to particle. Energy is transferred and spreads out through the material. This is conduction.

## Convection

Particles in liquids and gases have more energy than particles in a solid.

When heated, particles gain energy and move more rapidly. They move further away from one another. Therefore, a liquid or gas expands and becomes less dense. It rises into the cold areas and denser, cold liquid or gas falls into the warm areas. **Convection currents** are created which transfer energy from place to place.

> ## Remember
> Both conduction and convection use particles to transfer energy. Radiation, however, can transfer energy through a vacuum.

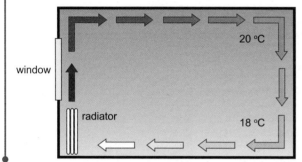

FIGURE 3: Air near the radiator is heated and moves around the room by convection currents.

> ## QUESTIONS
>
> **4** Describe how energy moves through a metal rod.
>
> **5** Describe what causes convection currents in a room.

# Infrared radiation

All objects **emit** infrared radiation. A hot object emits infrared radiation faster than a cooler one. The bigger the difference in temperature between an object and its surroundings, the faster energy is transferred. A Leslie's cube can be used to show that, in the same length of time:

> black surfaces give out more infrared radiation than white surfaces

> dull surfaces give out more infrared radiation than shiny surfaces.

> ## QUESTIONS
>
> **6** List the four surfaces of the Leslie's cube in the order that they emit infrared radiation, starting with the greatest rate.
>
> **7** Suggest why houses in very hot countries are often painted white.

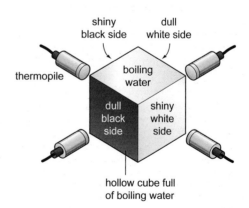

FIGURE 4: Using a Leslie's cube to compare infrared radiation from different surfaces.

# Insulation and payback time

**Essential notes**

- the U-value is the measure of the rate of heat loss through a material
- energy-saving measures such as insulation have a set-up cost
- payback time is how long it takes to repay set-up costs from energy savings made

**THE SCIENCE IN CONTEXT** Building regulations include regulations about energy efficiency – for example, the thickness of insulation required in buildings. The purpose is to reduce energy demands, costs and environmental effects. However, installing energy-saving measures also costs money. By comparing efficiency savings with installation costs, energy advisers can estimate 'payback times'.

## Saving energy for free

Some very rough energy-saving approximations: turning a room thermostat down by 1 °C could save £50 a year; not leaving electrical appliances on stand-by could save £35 per year; switching off lights when leaving a room could save £8 per year; only putting as much water in a kettle as is needed could save £5 a year. Can you think of other ways?

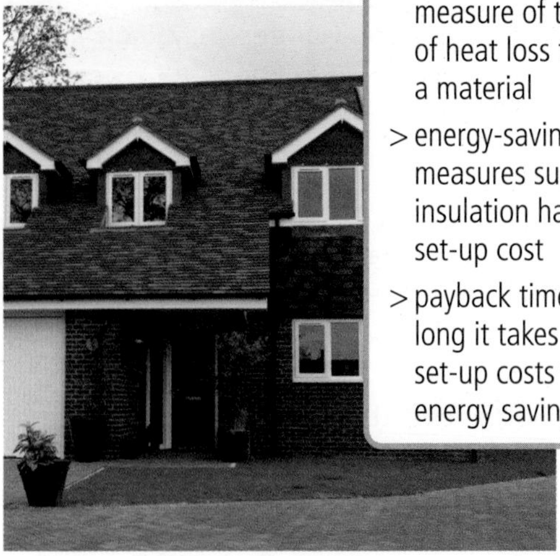

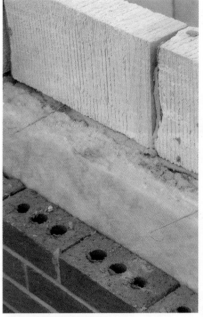

**FIGURE 1**: Heat loss happens through the roof, walls, windows, doors and floor.

## U-values

Using less energy for heating saves money and helps to protect the environment. Buildings are insulated to reduce the transfer of energy from them to their surroundings. Double glazing, cavity wall insulation and good loft insulation all reduce the amount of energy needed to keep homes warm.

**U-values** show how well a material acts as a thermal **insulator**. A low U-value means it is hard for energy to flow through a material, so the material is a good insulator. To insulate a loft, for example, a material with a low U-value is needed.

In many insulators, the material traps bubbles of air (or gas). It is this combination that makes them good insulators.

### Cavity walls

Before about 1920, most external walls were solid brick. After that, walls were usually built in two layers, with air between the bricks. These are called cavity walls. They provide better insulation than solid brick walls (see Table 1). Nowadays the cavity is filled with an insulating material. This reduces the possibility of convection currents.

| Material | Approximate U-value (W/m$^2$K) |
|---|---|
| solid wall | 2.2 |
| cavity wall (unfilled) | 1.0 |
| cavity wall (insulated) | 0.6 |

**TABLE 1**: U-values for external walls.

**FIGURE 2**: Insulation fills the cavity in this wall.

## Double glazing

Glass is a reasonable insulator. A single glazed window (single pane of glass) has a U-value of about 5.0 W/m²K. A double glazed window, with air or **argon** trapped between the two panes of glass, has a U-value of about 3.0 W/m²K. This is because air and argon, like all gases, are very good insulators.

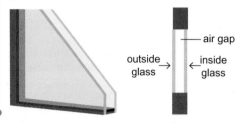

air gap

outside glass → ← inside glass

**FIGURE 3**: Glass and air together provide effective insulation.

### QUESTIONS

**1** Give reasons why it is a good idea to use less energy.

**2** Explain why cavity walls and double glazed windows have lower U-values than solid walls and single glazed windows.

## Payback time

By installing insulation, people in an average UK home could reduce their heating costs. They would reduce energy wasted through roofs, walls and windows. Unfortunately, the work and materials actually cost money – so is insulation worthwhile?

**Payback time** is the time that it takes to save as much money as the insulation cost to install.

| Method of insulation | Installation cost (£) | Annual saving (£) | Payback time (years) |
|---|---|---|---|
| loft insulation | 240 | 60 | 4 |
| cavity wall insulation | 360 | 60 | 6 |
| draught proofing doors and windows | 45 | 15 | 3 |
| double glazing | 2400 | 30 | 80 |

**TABLE 2**: Payback time for different insulation methods

Draught proofing doors and windows pay back the quickest (Table 2). Double glazing takes a lifetime to pay back, yet installing it is still very popular. Rather than to save money on fuel bills, people often install double glazing to reduce noise, from outside, or to save money on repainting wooden window frames.

**FIGURE 4**: Would you recommend that people install loft insulation?

### QUESTIONS

**3** Which type of insulation is cheapest to install?

**4** Describe what 'payback time' means.

**5** If draught proofing lasts five years before it needs replacing, and loft insulation lasts 10 years, suggest which you would install and why.

## Efficiency and cost-effectiveness

U-values are a measure of the efficiency of different insulating materials. They show how effectively a material insulates. This is not the same as cost-effectiveness.

Some forms of insulation, for example, may be efficient but expensive to buy and install. If the payback time is long, you might not think the measures taken were cost-effective (in other words, value for money).

### QUESTIONS

**6** Explain the difference between efficiency and cost-effectiveness.

# Preparing for assessment: Applying your knowledge

*To achieve a good grade in science, you not only have to know and understand scientific ideas, but you need to be able to apply them to other situations and investigations. These tasks will support you in developing these skills.*

## ☀ Eco-friendly makeovers

Most people recognise the need to reduce the carbon footprint of buildings, in other words, making them more energy efficient. New buildings are designed and built to standards that demand energy efficiency, but what about older buildings? One option is to demolish and replace them with new buildings. The other is to give them an eco-friendly makeover to increase their energy efficiency – much like the bungalow in the photograph.

The Sustainable Energy Academy (SEA) is spearheading a project called '*Old Home SuperHome*'. It is looking for older buildings that have been retrofitted with various measures to improve their energy efficiency. A network of homes that show good energy efficiency in practice is being established, with the goal that nearly everybody in the country will live near one and be able to visit and learn from it.

An example is a house in Powys, in Wales. When it was built 200 years ago, like many older houses it was not built with insulation in mind. However, following a retrofit it is claimed that carbon emissions have been cut by low energy lights, a wood pellet stove and a boiler.

The owner of the house in Powys also said that he had not kept track of how much it had cost him to refurbish his home. However, everything he did to increase energy efficiency (for example, fitting external insulation, secondary double glazing, solar thermal panels and low energy lights) cost money.

Similar to the house in Powys, this 1950s bungalow has, among other things, cavity wall insulation, roof insulation, triple glazing, solar PV panels and water heating, and a wind turbine.

## ✹ TASK 1

A carbon footprint is a measure of the amount of carbon dioxide and other greenhouse gases produced through burning fossil fuels – for example to generate electricity, provide heating and power vehicles.

(a) Name three greenhouse gases.

(b) Explain what activities might contribute to the carbon footprint of a home.

## ✹ TASK 2

(a) Suggest three ways which are now used to make newer houses more energy efficient than those built in the 1950s.

(b) The 'Old Home SuperHome' project surveyed people to find out why they decided to improve the energy efficiency of their homes. Suggest two reasons you would give.

## ✹ TASK 3

The owner of the house in Powys said, "New homes have cavity wall insulation, but older properties have solid walls. I've managed to externally insulate about 70% of my walls."

(a) Name and describe the three ways that energy is moved from place to place in a home.

(b) Heating engineers use U-values. State what a U-value is and what it means if a material has a high or low U-value.

(c) Explain why the solid walls in older buildings are not as effective at insulating a house as cavity walls. How can the effectiveness of a cavity wall be increased further?

## ✹ TASK 4

(a) Explain the term 'payback time'.

(b) Winnie recently retired. She lives alone in a small semi-detached bungalow. It gets very cold in the winter and is costly to heat. She has a limited amount of money to spend to make her bungalow easier to keep warm. Winnie is considering having the loft insulated, double glazing fitted and the doors and windows draught proofed. She cannot do it all though. What advice would you give her? Explain your reasoning.

## ✹ MAXIMISE YOUR GRADE

| Answer includes showing that you can... | |
|---|---|
| **E** | recall that carbon dioxide, methane, nitrous oxide are natural greenhouse gases. |
| | recall that burning fossil fuels and decomposing rubbish in landfill sites increase emissions of natural greenhouse gases. |
| | describe a U-value as a measure of the rate of heat loss through a material. |
| | suggest implications of being more energy efficient – for example, environmental issues and cost. |
| **C** | state and describe energy transfer by conduction, convection and infrared radiation. |
| | recognise and use data about ways of reducing domestic energy consumption and waste – for example, insulation and double glazing. |
| **A** | explain 'payback time' in terms of how long it takes to save as much money as the cost of 'energy efficiency' installations. |
| | discuss the cost-effectiveness of methods used to reduce domestic energy consumption. |

# Pollution indoors

**Essential notes**

> pollution-free air in the home is important for health and wellbeing

> indoor pollution can be more serious than outdoor pollution

**THE SCIENCE IN CONTEXT** You may have read about air pollution, but did you know it is also a problem indoors? In 2009, the World Health Organization (WHO) said that pollution caused by fungi (moulds) and bacteria is a key element of indoor air pollution. Scientists believe that, in some situations, indoor pollution can be a more serious problem than outdoor pollution.

## Second-hand smoking

You breathe in second-hand smoke when you are in the same room as a smoker, or where a smoker has been. Even though you may not be able to smell it, the smoke will still be there two and a half hours after opening a window. The 4000 toxic chemicals in the smoke put you at risk of the same smokers' diseases, including heart disease and cancer. No wonder there are now strict laws in force to keep public areas smoke free.

**FIGURE 1**: This mother and daughter are both at risk of heart disease and cancer.

## Pollution-free air

Air pollution is the presence of anything in the air that can cause ill-health in humans or animals or damage to the environment.

People spend a lot of time indoors, and environmental scientists have recognised that pollution-free air inside the home is important for health and wellbeing.

Many modern houses suffer from very poor air quality. In the push to become ever more energy efficient, ventilation is very poor. Windows and doors are kept closed and sealed to prevent draughts. The air is contained inside and not changed. Contaminants can build up so that air inside the home can often be more polluted than that outside.

Common pollutants include:

> dust

> mould and spores

> pollen

> smoke

> fumes from household products.

Modern homes and offices are often air conditioned. This may recycle the same air many times over and make indoor air pollution problems even worse. Poor air-conditioning systems can:

> circulate air – transmitting infectious respiratory diseases such as colds and flu

> circulate dust and moulds – causing allergic reactions, including asthma and itchy noses

> make the air too dry – causing sore throats and red eye (conjunctivitis).

**FIGURE 3**: These new windows have vents at the top. Why is it important to keep the vents open?

## QUESTIONS

**1** Explain what is meant by 'air pollution'.

**2** State the main cause of air pollution problems in the home.

**3** Describe the problems that air pollutants in houses are likely to cause.

# Problems with pollutants

Moulds and dust can be difficult to control, but they can be reduced by improving ventilation that will regularly change the air inside the home.

## Dust mites make you sneeze

Household dust contains a large number of tiny particles worn away from household items. It also includes shed human skin and large amounts of dust mite faeces. About 0.4 mm long, dust mites thrive in carpets, furniture, mattresses and bedding. They feed mainly on shed human skin. Around 100 000 can live in one square metre of carpet or bedding.

## More about moulds

Moulds are fungi. They reproduce by releasing millions of microscopic spores into the air. They grow best in moist conditions – so are most often found in damp, poorly ventilated areas such as bathrooms.

## Allergic reactions

Dust mite faeces and airborne mould spores can cause allergic reactions:

> hayfever-like symptoms – sneezing, a runny nose and itchy eyes

> asthma attacks – severe breathing difficulties in asthmatics.

The American Institute of Medicine found evidence to link indoor exposure to moulds with:

> respiratory symptoms, such as coughs and wheezes, in otherwise healthy people

> asthma symptoms in people who had asthma.

## Good air conditioning

Well maintained modern air-conditioning systems can include filters which help to remove dust and moulds from the air. Good ventilation to change polluted air for fresh air remains an excellent method for alleviating all problems.

FIGURE 3: Dust mite magnified 300 times. Many people are allergic to dust mite faeces.

### Did you know?

Asthma affects more than three million people in the UK.

### QUESTIONS

**4** Explain why dust mites are such a problem.

**5** Describe how moulds cause air pollution.

# Fumes from household products

Many household products can add to domestic air pollution. Volatile organic compounds (VOCs) are chemicals that **vaporise** easily at room temperature. Sources include paints, vinyl flooring and synthetic furnishings. Some, such as benzene, may cause cancers.

Recently, fragrances have also been called into question. Perfumes and air-fresheners do not list the complex cocktail of substances that they contain. Among other dubious ingredients often found is diethyl phthalate, linked to **sperm** damage.

### QUESTIONS

**6** Explain why the ingredients in fragrances are often kept secret.

**7** Describe precautions that you could take to protect yourself from VOCs.

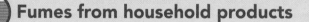

Q ... household fumes

# Household products

**THE SCIENCE IN CONTEXT** You might not think of kitchens or bathrooms as being dangerous places and, on the whole, they are not. However, many of the products used in homes are hazardous and should be used with care. You should follow the guidance given on their labels. They also need to be stored safely.

## Poisoning of children

Every year in the UK, over 28 000 children receive treatment for poisoning or suspected poisoning. Most incidents involve medicines, household products or cosmetics. Hazardous materials are often stored where children can find them, in containers which they can open.

**FIGURE 1**: Common household chemicals can pose a serious risk to children.

## Hazards and risks

A hazard is anything which can cause harm. A risk is the likelihood that a person will be harmed if exposed to a hazard. It is also usual to take into account the likely severity of the harm that might occur.

### Risk assessment

Risk assessment is the process in which you:

> identify hazards

> evaluate the risk associated with the hazard

> decide on suitable methods for eliminating or controlling the hazard.

### Risk management

Risk management is how you create a safe situation. The risks are assessed. Then health and safety measures are put in place to prevent harm or make it unlikely.

Any risk taken is according to ALARP: As Low As Reasonably Practicable.

### Did you know?

In England and Wales, 12 children are admitted to hospital every day with suspected poisoning.

### QUESTIONS

**1** Crossing the road can be hazardous. Explain how you can minimise the risk of injury when crossing a road.

**2** Describe how you would make it possible to use a hazardous product.

## Product labelling

### CHIP and CLP

Labelling requirements are changing. In the UK the CHIP (Chemicals – Hazard Information and Packaging for Supply) Regulations are being replaced by the European CLP (Classification, Labelling and Packaging of substances and mixtures). Square yellow hazard warning symbols are being replaced by diamond-shaped symbols with red borders.

There are also special requirements for packaging, including the use of child-resistant caps.

**FIGURE 2**: A CLP toxic warning symbol.

## Household product safety

At work, strict regulations deal with the use of hazardous substances and require risk assessments. You do not have to do this at home, which makes accidents more likely. What can you do to avoid accidents in the home with hazardous household products?

Information on the hazards posed by many household products is given on their labels. There is also advice on measures that you can take to reduce the risk.

This information can be used to make your own risk assessments, so that it is possible to:

> store and use household products safely

> decide which products are safe and which should be replaced by safer alternatives.

## Household cleaners

How much thought have you given to cleaning products as hazards?

Consider the risks associated with the use and storage of multi-surface cleaners.

A supermarket own brand cleaner has this information on its label:

**IRRITANT** *Keep out of the reach of children. Irritating to the eyes and skin. Avoid contact with skin and eyes. In case of contact with eyes, rinse immediately with plenty of water and seek medical advice. If swallowed, seek medical advice immediately and show this container or label. Contains Sodium Hypochlorite and Sodium Hydroxide. Warning! Do not use together with other products. May release dangerous gases (chlorine).*

Some household products are more hazardous. A well-known drain-clearing product, for example, carries the symbols:

***CORROSIVE, OXIDISING, HARMFUL* and *HIGHLY FLAMMABLE*.**

**FIGURE 3**: Even basic household cleaners can be hazardous and carry risk. They can cause harm to people if misused.

## QUESTIONS

**4** Find a household cleaning product. Look at its label and make a risk assessment. You should (a) identify any hazards (b) suggest action that can be taken to reduce the risk (c) decide if the risk is acceptable for the product to be used.

# Using less toxic products

Some people think the risks of using many of the available household products are unacceptable, especially if potential harm to the environment is taken into account.

| Cleaning product | Hazard | Non-toxic alternative |
|---|---|---|
| ammonia in glass cleaner | lung and skin irritant | diluted vinegar |
| furniture polish | nitrobenzene linked to cancers and birth defects | cornstarch |
| toilet cleaner | eye and skin irritant; naphthalene fumes can cause liver and kidney damage | lemon juice |
| surface cleaner | eye and skin irritant | sodium carbonate (washing soda) or bicarbonate |
| scented cleaner | limonene, for example, can damage liver, kidneys and bone marrow | |

## QUESTIONS

**5** Hold a class discussion. Do the advantages of using powerful commercial cleaning products outweigh the risks?

🔍 ... eco cleaning

# Boiler safety

**THE SCIENCE IN CONTEXT** In a home, water is usually heated up in a boiler. Many boilers are heated by burning fossil fuels. However, if the boiler is not working properly, the fossil fuel might not combust completely and poisonous carbon monoxide might be produced. This is why homes with gas boilers have carbon monoxide detectors near the boiler.

## Old boilers

Water heating boilers that are well maintained by properly trained and qualified fitters are safe and efficient. Unserviced boilers can become hazardous and pose an unacceptable risk to life. Landlords are legally required to have boilers checked by registered fitters every year. They must have a certificate to confirm that appliances have been inspected and passed as safe.

**FIGURE 1**: A domestic boiler. Safe if well maintained. Potentially deadly if not.

## Boiler efficiency

To convert all the energy stored in a fuel into heat it must be burned completely. Wood and fossil fuels such as gas, coal, coke and oil contain carbon. Complete combustion gives carbon dioxide and the maximum amount of heat:

$$C + O_2 \rightarrow CO_2 + heat$$

Fuels may also contain hydrogen, which is converted to water when the fuel is burned.

To burn completely, any fuel needs an adequate supply of air to provide enough oxygen. Incomplete combustion causes some of the carbon to be converted to carbon monoxide and some to remain as carbon particles.

> Carbon monoxide is an extremely poisonous gas.

> Unburned carbon forms smoke and soot.

### Did you know?

Early boilers were dangerous and sometimes exploded, causing loss of life. The Institution of Chartered Mechanical Engineers was founded in 1847 by George Stephenson to provide expert knowledge and improve safety.

**FIGURE 2**: Fuel was not used efficiently in this old boiler.

## QUESTIONS

**1** What, in air, is needed for burning fuels?

**2** What will happen to the temperature of water leaving a boiler if there is an insufficient air supply?

**3** Why is an inadequate supply of air to a boiler dangerous?

# Bunsen burner model

The air hole at the base of a Bunsen burner allows air to mix with gas before it burns.

The burner should be lit with the air hole closed. The flame will then change as the air hole is opened.

**Combustion** of **fuel** in a domestic boiler will also be affected in the same way, depending on the amount of air that reaches the burning fuel.

When fuel burns in an enclosed room, the amount of oxygen gradually decreases and the amount of carbon dioxide increases. This causes a switch from complete combustion to incomplete combustion, and the production of increasing amounts of carbon monoxide (CO).

FIGURE 3: On each of these Bunsen burners, are the air holes open or closed?

## QUESTIONS

**4** For a Bunsen burner: (a) What is the appearance of the flame with the air hole closed? (b) Suggest what is present in the flame that gives the characteristic yellow colour. (c) Suggest how you could demonstrate that your answer to (b) is correct. (d) Why is the flame different to (a) when the air hole is fully open? (e) Which flame would heat a beaker of water faster, and why?

**5** (a) Suggest methods that you could use to check if a boiler was not receiving a sufficient supply of air. (b) What might cause a boiler to have an insufficient air supply? (c) How could you improve the air supply to a boiler?

# Carbon monoxide poisoning

Up to 50 people per year, in the UK, are killed by carbon monoxide poisoning. This makes it the most common cause of fatal poisonings in the UK. Many more people are poisoned by it, but survive.

Carbon monoxide is:

> a colourless, odourless and non-irritant gas

> produced by incomplete combustion of carbon-containing fuels

> produced in poorly maintained, unserviced or badly ventilated heaters.

## How it kills

Oxygen is mainly transported in the blood as oxyhaemoglobin. This is formed when haemoglobin, the red pigment in red blood cells, takes up oxygen in the lungs. It then releases it in the tissues, where it is needed for respiration.

Carbon monoxide can also bind to haemoglobin. It does so in preference to oxygen and binds tightly to form carboxyhaemoglobin. This progressively reduces the amount of oxygen that can be carried by the blood.

Affected people suffer from headaches, dizziness, nausea and weakness. Eventually, they lapse into unconsciousness and die when insufficient oxygen reaches their brain.

## QUESTIONS

**6** (a) Why is carbon monoxide so dangerous? (b) What precautions can be taken to prevent carbon monoxide poisoning?

# Background radiation

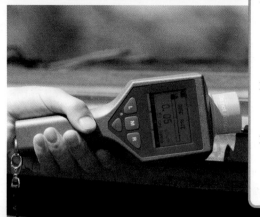

**THE SCIENCE IN CONTEXT** It may surprise some people to learn that they live on radioactive land. The cause is radioactive radon gas. Before panic sets in, however, the Health and Safety Executive (HSE) says that radon hazards are simple and cheap to measure and, if levels are high, relatively easy to address. The Health Protection Agency lists accredited laboratories that can carry out the measurements.

## Essential notes

> radium and uranium occur naturally in the ground and emit background radiation in radon gas

> radon gas is radioactive and a cause of cancer

> radon may accumulate beneath homes and become a pollutant

### High levels of radiation

In 2004 the National Radiological Protection Board found two homes in Cornwall with radon concentrations 85 times higher than the recognised safety limits.

**FIGURE 1**: Geiger counters can be used to detect radiation levels.

## Background radiation

Radiation, as particles or rays, can be detected using a Geiger counter. If a Geiger counter is switched on, there will be a low level of clicks from the counter. This shows that there is a small amount of radiation around us all the time. This is background radiation. It comes from a wide range of sources. Some is from space, some from sources made by people, and some is from the soil and rocks on the Earth.

This background radiation is at such a low level that it is thought not to be harmful. Our bodies have evolved mechanisms that repair any damage that it may cause.

### The ground beneath our homes

Surveyors have realised that radioactive pollution may be caused by the types of soil beneath our homes. Radon may become a pollutant in the air in houses if the rocks and soil underneath a house contain large enough concentrations of radium or uranium.

This is especially the case in areas of the UK where granite rock is found, such as Scotland and Cornwall.

Radon is a colourless, odourless, radioactive gas which is a cause of cancer, especially lung cancer.

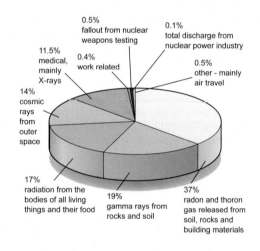

**FIGURE 3**: The sources of background radiation.

### QUESTIONS

**1** Explain why a working Geiger counter never gives a zero reading.

**2** Where is radon most likely to be an air pollutant inside homes?

**3** Explain why radon is hazardous.

Q ... gcse background radiation

# The risk from radon

Because it is a gas, radon can be breathed into the lungs. As radon decays it emits alpha particles. These large, positively charged particles are powerful ionising radiation. They have low penetrating power but, once inside the lungs, they can cause significant damage.

Radon is the second most prevalent cause of lung cancer after smoking. It is the main cause of lung cancer in non-smokers.

## Radon in houses

For most people, the greatest exposure to radon is in their home. The concentration of radon in a house depends on:

> how much uranium there is in the soil and rocks under the house

> how the radon is able to enter the house

> how well the house is ventilated.

Radon is able to enter houses through:

> gaps in floors

> gaps between **concrete** floors and walls

> drains.

## Reducing radon levels

Radon levels in houses can be reduced by:

> sealing floors

> improving house ventilation

> increasing under floor ventilation (non-solid floors).

New houses built in high-risk areas include protection, such as solid floors, to keep levels down.

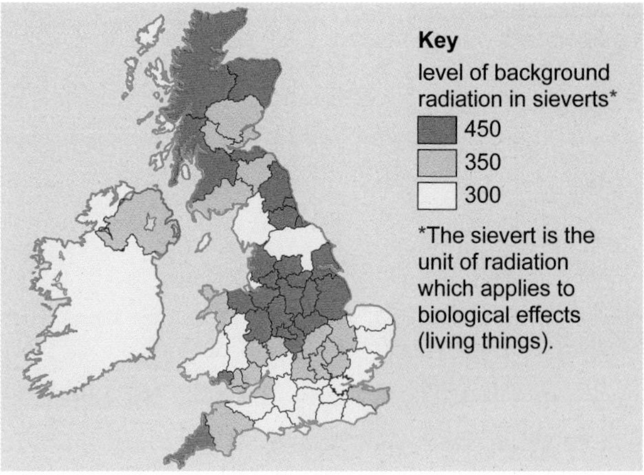

**Key**

level of background radiation in sieverts*

- 450
- 350
- 300

*The sievert is the unit of radiation which applies to biological effects (living things).

**FIGURE 4**: Background radiation in the UK.

## QUESTIONS

**4** Describe what makes radon especially dangerous.

**5** Describe how radon enters houses.

**6** Explain how the various safety measures taken can reduce exposure to radon.

# Smokers' risk

In 2004 scientists researching the incidence of radon-induced lung cancer looked at the risks to smokers and non-smokers. Their findings are summarised in Table 1. Radioactivity is measured in becquerels per cubic metre (Bq per $m^3$).

| Exposure to radon (Bq per $m^3$) | Risk of death from lung cancer (%) | |
|---|---|---|
| | Non-smokers | Smokers |
| 0 | 0.4 | 10 |
| 100 | 0.5 | 12 |
| 400 | 0.7 | 16 |

**TABLE 1**: Risk of lung cancer for smokers and non-smokers exposed to radon.

### Did you know?

In the UK, radon in the home causes about 1000 deaths per year from lung cancer.

## QUESTIONS

**7** By approximately how much does the risk of death from lung cancer differ between smokers and non-smokers?

# Unit 3 Theme 3 Checklist

## To achieve your forecast grade in the exam you will need to revise

Use this checklist to see what you can do now. Refer back to the relevant pages in this book if you are not sure. Look across the three columns to see how you can progress. **Bold** text means Higher tier only.

Remember that you will need to be able to use these ideas in various ways, such as:

> interpreting pictures, diagrams and graphs

> applying ideas to new situations

> explaining ethical implications

> suggesting some benefits and risks to society

> drawing conclusions from evidence that you are given.

Look at pages 250–271 for more information about exams and how you will be assessed.

| To aim for a grade E | To aim for a grade C | To aim for a grade A |
| --- | --- | --- |
| Know that making and using products may result in increased emissions of greenhouse gases into the atmosphere. | Describe the main ways in which making and using products may result in increased emissions of natural greenhouse gases into the atmosphere. | Explain how human activity results in the production of carbon dioxide, methane and nitrous oxide. |
| Know that greenhouse gases may cause global warming. | Explain how increased greenhouse gases absorb more long-wave radiation from the Earth and retain heat in the atmosphere. | Know about international agreements such as the Kyoto agreement on climate change to achieve the stabilisation of the dangerous gases carbon dioxide, methane and nitrous oxide. |
| Recall that leaching of artificial fertilisers, pesticides and herbicides causes pollution in lakes and rivers. | Explain the process of eutrophication resulting from overuse of fertilisers. | Explain how indicator species such as freshwater invertebrates and lichens may be used to monitor changes in pollution levels. |
| Recall that wate may be disposed of by burial in landfill sites, incineration or recycling. | Describe the environmental impact of landfill sites for the disposal of waste materials including plastics. | **Understand that non-degradable products use productive land in landfill.** **Understand that partial breakdown produces toxic materials that may leak into the environment.** |

## To aim for a grade E    To aim for a grade C    To aim for a grade A

| To aim for a grade E | To aim for a grade C | To aim for a grade A |
|---|---|---|
| Recall that plants may be used to make plastics. | Describe advantages and disadvantages of using plants to make plastics.<br><br>Describe advantages and disadvantages of using biodegradable products in landfill. | **Know that biodegradable products break down into substances that may be useful.**<br><br>**Describe the methods of degrading plastics as photo-degradable or oxo-degradable.**<br><br>**Explain why plastics such as polyvinyl alcohol (PVOH) and ethylene vinyl alcohol (EVOH) are used for plastic films for packaging and shopping bags.** |
| Describe how heat is transferred by conduction, convection and radiation in the home. | Recall that the U-value is the measure of the rate of heat loss through a material.<br><br>Describe ways of minimising heat loss in the home. | Understand how to interpret U-value data. |
| Explain the difference between efficiency and cost-effectiveness. | Describe the efficiency and cost-effectiveness of methods used to reduce domestic energy consumption. | Explain the term 'payback time' in relation to installing energy-saving measures. |
| Know that environmental scientists recognise that pollution-free air in the home is important for our health and wellbeing. | Name common pollutants in homes.<br><br>Name common symptoms of exposure to high indoor pollution levels. | Describe the importance of ventilation in the home. |
| Recall hazards caused by using household products. | Interpret hazard labels on household products.<br><br>Recall risks associated with household product hazards, and describe ways of minimising these risks. | Describe methods of reducing pollution in the home, including the use of less toxic products. |
| Explain why domestic boilers need an adequate supply of air to work efficiently. | Explain how incomplete combustion of fuels used in domestic boilers results in lower energy output and the formation of toxic combustion products and soot. | |
| Recall that radon is a radioactive gas and is a cause of cancer. | Understand that if rocks and soil beneath the home contain large concentrations of radium or uranium, radon may become a pollutant. | Describe methods to reduce radon levels in homes. |

# Unit 3 Exam-style questions: Foundation

AO1 **1.** Pain relief is a popular topic in advertising. Statements advertising headache tablets are often seen.

*Faster and longer relief from headaches*
*Action on the pain is twice as fast as with other medicines*

Is the statement, 'My headache is cured by taking paracetamol', true or false? Give a reason for your answer. [2]

**2.** There is concern about the increased use of recreational drugs taken by young people. Smoking and alcohol abuse are topics always in the news.

AO1 **(a)** Name two recreational drugs, other than tobacco and alcohol, that may harm the body. [2]

AO1 **(b)** People who become dependent on drugs usually suffer withdrawal symptoms when they stop using them. Why does this happen? [1]

AO2 **(c)** 'You can eat five portions of fruit and veg a day and exercise regularly, but healthy behaviour means little if you continue to smoke.' Give two reasons why this is good advice. [2]

AO2 **(d)** Nearly everybody is familiar with the old message that 'smoking is bad for you'. Tobacco smoke contains substances that cause disease. Name three substances found in tobacco that can cause harm. [3]

**3.** A student was asked to make a presentation on the link between smoking and lung disease. He found the following data from Cancer Research UK on the internet.

| Age group | Percentage of people smoking | | | | | | | |
|---|---|---|---|---|---|---|---|---|
| | 1974 | 1986 | 1994 | 2004 | 2005 | 2007 | 2008 | 2009 |
| 16–19 | 40 | 30 | 27 | 24 | 24 | 21 | 22 | 24 |
| 20–24 | 48 | 39 | 39 | 32 | 32 | 31 | 30 | 26 |
| 25–34 | 51 | 36 | 32 | 31 | 31 | 26 | 27 | 25 |
| 35–49 | 52 | 36 | 30 | 29 | 27 | 24 | 24 | 25 |
| 50–59 | 51 | 35 | 27 | 24 | 24 | 21 | 22 | 21 |
| 60+ | 34 | 25 | 17 | 14 | 14 | 12 | 13 | 14 |

| Age range | 25–34 | 35–44 | 45–54 | 55–64 | 65–74 |
|---|---|---|---|---|---|
| Number of male deaths in 2008 | 10 | 137 | 845 | 3416 | 6293 |
| Number of female deaths in 2008 | 6 | 133 | 785 | 2559 | 4318 |

AO3 **(a)** Describe four trends supported by the data that he could put in his presentation. [4]

AO2 **(b)** There is now a ban on smoking in public buildings. Some people do not agree with this ban and are campaigning to have it lifted. Suggest one scientific reason to support the ban and explain your answer. [2]

**4.** *In this question you will be assessed on using good English, organising information clearly and using specialist terms where appropriate.*

Binge drinking is a common problem affecting many young people. Alcohol is not classed as an illegal drug but, as well as being addictive, it can cause long-term damage.

AO2 **(a)** Discuss the short- and long-term problems associated with binge drinking. [4]

AO2 **(b)** Suggest two ways to help reduce binge drinking in the UK. [2]

**5.** Medical professionals involved in the diagnosis and treatment of disease may use electromagnetic radiation in their work.

AO1 **(a)** Give the difference between diagnosis and treatment. [2]

AO1 **(b)** Name two types of electromagnetic radiation which can be used by the medical professionals. [2]

**6.** The use of electromagnetic radiation can be dangerous. Professionals working in a busy radiography department need to be protected from the harmful effects of radiation that they use in their daily work. The radiation can be monitored by using film badges.

AO1 **(a)** Describe the structure of a film badge. [2]

AO1 **(b)** How is the information that these badges provide used to protect the radiographers? [2]

**7.** Electroplating can be used to put a thin layer of copper on to another metal when making jewellery.

AO1 **(a)** Other than copper, name a metal usually used to electroplate nickel jewellery. [1]

AO1 **(b)** Give a reason why this metal is used in jewellery design. [1]

AO1 recall the science    AO2 apply your knowledge    AO3 evaluate and analyse the evidence

**8.** A student decides to set up an experiment to copper plate a bracelet. She is given a list of equipment and chemicals to use. This includes a power supply, two leads and crocodile clips, a strip of copper, a bracelet and a beaker containing copper sulfate solution.

AO1 **(a)** Complete the following sentence, choosing the correct word from this list to describe how the chemicals are used:

**solution, electrolyte, electrode, anode, cathode, ion, electrolysis**.

*Copper sulfate* _____ *is the* _____ , *the strip of copper is the* _____ *and the bracelet is placed as the* _____. [4]

AO1 **(b)** Draw a diagram to show how the student would put together the equipment so that she could electroplate the bracelet. [5]

AO2 **(c)** Another student decided that he wanted to copper plate his own bracelet but this was made of rubber. Give one reason why this would not work. [1]

AO3 **9.** *In this question you will be assessed on using good English, organising information clearly and using specialist terms where appropriate.*

Cotton is a natural material. It is cool to wear, very absorbent and dries slowly. It can be washed, but needs to be ironed. It creases easily.

*Lycra®* is a synthetic product, cheaper to produce than cotton. Among other things, it is used to make sportswear. *Lycra®* is crease-resistant, stretchy and 'non-iron'. It is absorbent and dries quickly.

Use the information above and your knowledge and understanding to compare the advantages and disadvantages of cotton and *Lycra®*. [6]

**10.** *'See your happy face develop as you drink your morning coffee'*

This is an advertisement that may be seen in photo shops. It is a photograph printed on a mug which can be seen developing as hot coffee is poured into the mugs.

AO1 **(a)** Give the general name given to the paint used to make these mugs. [1]

AO2 **(b)** Explain what happens to the paint. [2]

AO2 **(c)** Give another example of where these types of materials can be used. [1]

**11.** A student completing a quiz on gases needed to choose the gases which matched the given statements.

**oxygen, carbon dioxide, nitrous oxide, methane, nitrogen, hydrogen, air**

Give your choice for these quiz questions:

AO1 **(a)** Name the gas which is a hydrocarbon formed from the decomposition of rubbish. [1]

AO1 **(b)** Name the gas formed as a result of increased use of nitrogenous fertilisers. [1]

AO1 **(c)** Name the gas formed by the burning of fossil fuels. [1]

AO1 **(d)** Name the gas which is an element needed for combustion. [1]

**12.** Biological pollutants can travel through the air and can damage surfaces inside your home.

AO1 **(a)** Name two biological pollutants that may be found inside the home. [2]

AO1 **(b)** What symptoms may people complain of, if there are high levels of biological pollutants in the home? [2]

AO1 **(c)** It is important to read instructions for use carefully and to understand the meaning of hazard labels on cleaning products, for example, before using them.

State the hazard that the user has to be aware of on the following labels. [5]

**1.** We are constantly being invaded by microorganisms, particularly bacteria and viruses, which enter the body.

AO1 **(a)** Name two diseases caused by bacteria. [2]

AO1 **(b)** Name two diseases caused by viruses. [2]

AO2 **(c)** Explain why a person with an infectious disease needs to be kept away from other people. [2]

**2.** A student was asked to give a presentation entitled 'Immunity'.

AO1 **(a)** State what you understand by the term 'immunity'. [2]

AO2 **(b)** While carrying out her research she found the word 'pathogen'. Explain what this means. [1]

AO1 **(c)** One of her slides gave this information:

> The body's defences against disease can be divided into:
> • those which prevent the entry of the microorganisms, and
> • those which destroy them once they manage to get in.

State how the body blocks the entry of bacteria. Use the following natural barriers to help: **skin, sweat, tears**. [3]

AO2 **(d)** Describe four steps to show how white blood cells deal with an infection caused by a microorganism. Use the following words to help you: **antibodies, surface antigens, microorganism**. [3]

AO1 **(e)** The student found the words leucocytes, phagocytes and lymphocytes when carrying out her research and she wanted to use these. Give a meaning for each of these terms. [3]

**3.** YBCO is a ceramic compound of yttrium barium copper oxide. It is a superconductor at low temperatures.

AO1 **(a)** What are the advantages of using superconductors in terms of heat transfer in wires? [2]

AO2 **(b)** What are the advantages of using superconductors in MRI scanners? [2]

AO2 **(c)** Much research and development is taking place on superconductors. Suggest what the researchers are trying to develop. [1]

**4.** Antibiotics are powerful medicines that fight bacterial infections and if used properly, may save lives. They either kill bacteria or stop them reproducing. The following information was found on a medical website.

> After more than 50 years of widespread use, many antibiotics are losing their effectiveness. If something stops the ability of a microbe to spread, such as the use of antibiotics, genetic changes can occur that enable it to survive.
>
> Patients that do not complete a course of antibiotics can develop bacterial resistance.
>
> Patients with high risk of infection often need higher doses of antibiotics which can lead to selection of antibiotic-resistant microbes.
>
> Close contact among sick patients creates a fertile environment for the spread of antibiotic-resistant microbes.
>
> Scientists also believe that the practice of adding antibiotics to agricultural feed promotes drug resistance.

AO2 **(a)** Draw a labelled diagram to show how mutation can cause drug resistance. [4]

AO3 **(b)** *In this question you will be assessed on using good English, organising information clearly and using specialist terms where appropriate.*

Use your knowledge and the information above to discuss why the misuse of antibiotics results in bacterial resistance, which leads to an ongoing problem in healthcare. [6]

**5.** Electrolysis can be used to electroplate metals. In the electrolysis, for example, of copper sulfate solution using copper electrodes, copper ions are attracted to the cathode and copper metal is formed.

AO1 **(a)** Complete the equation given below to show the formation of copper metal.

$$Cu^{2+} + \_\_\_\_ \rightarrow Cu$$ [2]

AO1 **(b)** Write the equation to show what happens to the copper anode. [2]

**6.** Selective breeding is the traditional method for increasing disease resistance in plant crops, e.g. wheat or improving milk yields in cattle.

AO1 **(a)** State four steps to show how selective breeding can be used to increase disease resistance in a crop plant. [4]

AO2 **(b)** Describe the problems that might arise in future generations of cattle if selective breeding carries on unchecked. [4]

**7.** Genetic engineering is a fast way of producing plants or animals with the characteristics that we require. Human insulin can be made both quickly and relatively cheaply by using such a technique.

AO1 **(a)** Draw a series of labelled sketches to describe the production of human insulin. [6]

AO2 **(b)** *In this question you will be assessed on using good English, organising information clearly and using specialist terms where appropriate.*

There are many benefits to genetic engineering, but there are also risks.

Suggest some advantages and disadvantages of genetic engineering. [6]

**8.** The expectation for high-quality food has lead to the use of fertilisers, pesticides and herbicides.

AO2 **(a)** Describe how fertilisers can increase crop production. [2]

AO2 **(b)** Suggest how herbicides and pesticides can lead to increased yields of crops. [2]

AO1 **(c)** Although fertilisers can increase crop yields, they can cause pollution in lakes and rivers by a process known as eutrophication.

Explain this process of eutrophication. [4]

**9.** Polymers are extensively used in the manufacture of a wide range of packaging. Scientists are increasingly involved in the development of polymers that are biodegradable.

AO2 **(a)** Describe what happens to plastic which decomposes in this way. [1]

AO2 **(b)** How does this help to reduce environmental damage? [2]

AO1 **(c)** *'Bio-degradable dog-mess bags – useful for dog walkers and the environment.'*

Oxo-degradable plastics are said to be 'totally degradable'. Oxo-biodegradation is a two-stage process.

Describe the stages of oxo-biodegradation. [4]

**10.** A European study, funded by Cancer Research UK and the European Commission, showed these data:

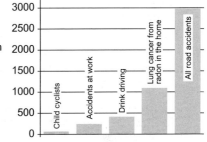

Radon deaths compared with other premature deaths annually in the UK

AO1 **(a)** Explain why radon is a risk to health. [2]

AO2 **(b)** Suggest where radon can be found other than in the home. [2]

AO2 **(c)** Suggest reasons why it is important to know the levels of radon in our homes. [2]

AO3 **(d)** Suggest why we should not be excessively worried about radon levels in our homes. [2]

AO1 recall the science   AO2 apply your knowledge   AO3 evaluate and analyse the evidence

## ✳ WORKED EXAMPLE – Foundation tier

Antibiotics are one of the most frequently prescribed medications.

**(a)** Name two infections that can be treated using an antibiotic. [2]

> *tooth abscess*
> *ear infection*

**(b)** Describe how antibiotics cure disease. [2]

> *By killing or injuring bacteria*

**(c)** Why do GPs not prescribe antibiotics to cure the common cold? [2]

> *Common cold is a virus and these are not affected by antibiotic.*

**(d)** Most antibiotics have two names, a brand name created by the drug companies and a generic name based on the chemical structure of the antibiotic. Give the name of an antibiotic based on the chemical structure. [1]

> *Penicillin*

**(c)** Many drugs have side effects. Explain how drug companies check the safety of their drugs before they are put on sale to the public. [4]

> *Laboratory research and extensive clinical trials*

**(d)** Many people are against the use of animals in drug testing. State one advantage and one disadvantage of this type of testing. [2]

> *Advantage – humans will not be harmed.*
> *Disadvantage – animals could be harmed*

### How to raise your grade!
Take note of these comments – they will help you to raise your grade.

Tooth abscess is a good example. Ear infections – take care when using this as an example as there are several types of ear infections. Antibiotics are used for some (but not all) ear infections, as some inner ear infections are caused by a virus.

Sinus infections – are treated using antibiotics.

Sore throat – again take care as most sore throats are caused by viruses; however, strep throat is an infection caused by bacteria. It is called 'strep' because the bacteria that causes the infection is called *Streptococcus*.

The answer is correct. Antibiotics work against infections caused by most bacteria. Yes – antibiotics work by killing the bacteria or by stopping them from growing.

This answer receives one mark for virus and one mark for knowing that viruses are not affected by antibiotics. Remember that antibiotics do not fight infections caused by viruses, such as colds, flu, most coughs and bronchitis, sore throats.

The answer is correct. There are many types – you could give a specific example, but do not give the brand name.

The answer is too brief, and will receive only two marks. If the question asks for an explanation, then there needs to be more than a simple statement.

The candidate could support the answer by explaining what is meant by a clinical trial. Clinical trials are research studies that test how well new medicines work on people; they use a range of volunteers. All data from these trials is analysed and checked to ensure safety, before any drugs can be used by the public.

The answer is correct. The question asks the candidate to state an advantage and disadvantage; there is no need to go into detail.

# ✴ WORKED EXAMPLE – Higher tier

*In this question you will be assessed on using good English, organising information clearly and using specialist terms where appropriate.*

In the past 50 years, vaccination has saved more lives than any other procedure or medical product. In 1980, smallpox was declared eradicated from the world and in 2008 scientists developed a new vaccine for cervical cancer.

In the first four months of 2009, there were more cases of mumps than in the whole of the previous two years. Teenagers and adults in their early 20s are at higher risk of mumps because many were too old to be routinely vaccinated with the Measles, Mumps and Rubella (MMR) vaccine.

The NHS offers a vaccination schedule to all ages. Flu (influenza) virus is spread by coughs and sneezes. Anyone can get flu but it can be more serious for 'high risk' people, including those over 65, pregnant women and those with a serious medical condition. These people are advised to have the 'flu jab'.

**(a)** Explain how vaccination can protect humans from infection against mumps. [6]

*A vaccination is an injection of weak or dead micro-organisms. This vaccine causes the body to produce antibodies which will kill the germs. An illness is due to micro-organisms causing you harm before your immune system can destroy them. If you become infected with a micro organism you have been vaccinated against you will have the specific antibodies in your blood before you get the infection and so you will not get ill.*

**(b)** Suggest why it is recommended for a 66 year old to have a 'flu jab' every year. [3]

*Protection only lasts for a limited time and therefore it is necessary to have further doses of the vaccine to keep up the protection – this is called a booster*

**(c)** Suggest why there is some concern about a 'vaccination schedule' and what implications this may have if younger people are not vaccinated. [3]

*A vaccination schedule is an effective way of controlling the spread of disease. There is risk when the micro organism /pathogen is injected in to the people but this is rare.*

*Some people are scared as they don't understand how a vaccine works – could improve health education*

*Some people just follow their friends and don't really understand the implications.*

## How to raise your grade!

Take note of these comments – they will help you to raise your grade.

The candidate's answer does explain what a vaccine is and how it works. However, it does not relate specifically to mumps. A statement that infection is caused by the mumps virus is important and that to immunise against viral diseases, the virus used in the vaccine has been weakened or killed.

Also, a statement about the immune system could be made, such as vaccinations work by stimulating the immune system (the natural disease-fighting system of the body). A healthy immune system can recognise invading viruses and produce substances (antibodies) to destroy or disable them.

The candidate has answered the question too generally. There is no reference to the 66 year old, so a mark would be lost. The 66 year old is in the high risk category. Protection could be lost after one year, so a fresh dose each year is needed.

The candidate has not really given an answer to the second part of the question – the implications if younger people are not vaccinated. The answer should include the idea that if people are not vaccinated, then they would once again develop such diseases. Epidemics could occur.

Also, if people do not accept vaccinations at the scheduled time, they could miss out completely.

# Carrying out practical investigations in GCSE Science

## Introduction

*As part your GCSE Science course, you will develop practical skills and have to carry out investigative work in science.*

Your investigative work will be divided into several parts:

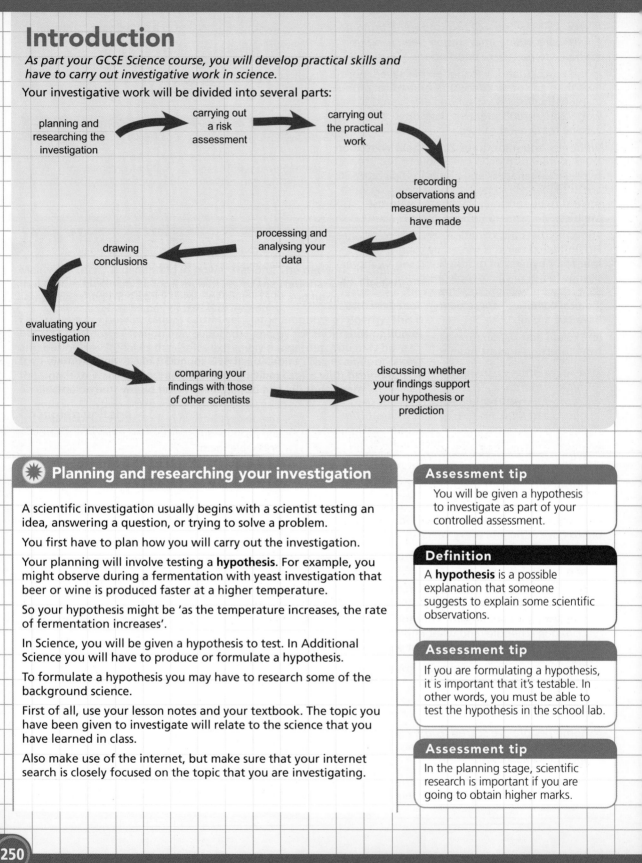

planning and researching the investigation → carrying out a risk assessment → carrying out the practical work → recording observations and measurements you have made → processing and analysing your data → drawing conclusions → evaluating your investigation → comparing your findings with those of other scientists → discussing whether your findings support your hypothesis or prediction

## ✸ Planning and researching your investigation

A scientific investigation usually begins with a scientist testing an idea, answering a question, or trying to solve a problem.

You first have to plan how you will carry out the investigation.

Your planning will involve testing a **hypothesis**. For example, you might observe during a fermentation with yeast investigation that beer or wine is produced faster at a higher temperature.

So your hypothesis might be 'as the temperature increases, the rate of fermentation increases'.

In Science, you will be given a hypothesis to test. In Additional Science you will have to produce or formulate a hypothesis.

To formulate a hypothesis you may have to research some of the background science.

First of all, use your lesson notes and your textbook. The topic you have been given to investigate will relate to the science that you have learned in class.

Also make use of the internet, but make sure that your internet search is closely focused on the topic that you are investigating.

### Assessment tip

You will be given a hypothesis to investigate as part of your controlled assessment.

### Definition

A **hypothesis** is a possible explanation that someone suggests to explain some scientific observations.

### Assessment tip

If you are formulating a hypothesis, it is important that it's testable. In other words, you must be able to test the hypothesis in the school lab.

### Assessment tip

In the planning stage, scientific research is important if you are going to obtain higher marks.

✔ The search terms you use on the internet are very important. 'Investigating fermentation' is a better search term than just 'fermentation', as it is more likely to provide links to websites that are more relevant to your investigation.

✔ The information on websites also varies in its reliability. Free encyclopaedias often contain information that hasn't been written by experts. Some question and answer websites might appear to give you the exact answer to your question, but be aware that they may sometimes be incorrect.

✔ Most GCSE Science websites are more reliable but, if in doubt, use other information sources to verify the information.

As a result of your research, you may be able to extend your hypothesis and make a **prediction** that's based on science.

> ### Example 1
>
> Investigation: Plan and research an investigation into the activity of enzymes
>
> Your hypothesis might be 'when I increase the temperature, the rate of reaction increases'.
>
> You may be able to add more detail, 'this is because as I increase the temperature, the frequency of collisions between the enzyme and the reactant increases'.

### Assessment tip

Make sure that you make a detailed note of which sources you have used: for a book, the author's name and the title; for a website, the name of it.

### Assessment tip

**Higher tier**
You are expected to be able to balance chemical equations. So, for example, if the enzyme is being used to decompose hydrogen peroxide to water and oxygen you should be able to balance the equation: $H_2O_2 \rightarrow H_2O + O_2$ to give $2H_2O_2 \rightarrow 2H_2O + O_2$

## ☀ Choosing a method and suitable apparatus

As part of your planning, you must choose a suitable way of carrying out the investigation.

You will have to choose suitable techniques, equipment and technology, if this is appropriate. How do you make this choice?

You will have already carried out the techniques you need to use during the course of practical work in class (although you may need to modify these to fit in with the context of your investigation). For most of the experimental work you do, there will be a choice of techniques available. You must select the technique:

✔ that is most appropriate to the context of your investigation, and

✔ that will enable you to collect valid data, for example if you are measuring the effects of light intensity on photosynthesis, you may decide to use an LED (light-emitting diode) at different distances from the plant, rather than a light bulb. The light bulb produces more heat, and temperature is another independent variable in photosynthesis.

Your choice of equipment, too, will be influenced by the measurements that you need to make. For example:

✔ you might use a one-mark or graduated pipette to measure out the volume of liquid for a titration, but

✔ you may use a measuring cylinder or beaker when adding a volume of acid to a reaction mixture, so that the volume of acid is in excess to that required to dissolve, for example, the calcium carbonate.

### Assessment tip

Technology, such as data-logging and other measuring and monitoring techniques, for example heart sensors, may help you to carry out your experiment.

### Definition

The **resolution** of the equipment refers to the smallest change in a value that can be detected using a particular technique.

### Assessment tip

Carrying out a preliminary investigation, along with the necessary research, may help you to select the appropriate technique to use.

## ✸ Variables

In your investigation, you will work with independent and dependent variables.

The factors you choose, or are given, to investigate the effect of are called **independent variables**.

What you choose to measure, as affected by the independent variable, is called the **dependent variable**.

## ✸ Independent variables

In your practical work, you will be provided with an independent variable to test, or will have to choose one – or more – of these to test. Some examples are given in the table.

| Investigation | Possible independent variables to test |
|---|---|
| activity of yeast | > temperature<br>> sugar concentration |
| rate of a chemical reaction | > temperature<br>> concentration of reactants |
| stopping distance of a moving object | > speed of the object<br>> the surface on which it's moving |

Independent variables can be **categoric** or **continuous**.

> When you are testing the effect of different disinfectants on bacteria you are looking at categoric variables.
> When you are testing the effect of a range of concentrations of the same disinfectant on the growth of bacteria you are looking at continuous variables.

### Range

When working with an independent variable, you need to choose an appropriate **range** over which to investigate the variable.

You need to decide:

✔ which treatments you will test, and/or

✔ the upper and lower limits of the independent variables to investigate, if the variable is continuous.

Once you have defined the range to be tested, you also need to decide the appropriate intervals at which you will make measurements.

The range you would test depends on:

✔ the nature of the test

✔ the context in which it is given

✔ practical considerations

✔ common sense.

**Definition**

Variables that fall into a range of separate types are called **categoric variables**.

**Definition**

Variables that have a continuous range are called **continuous variables**.

**Definition**

The **range** defines the extent of the independent variables being tested.

## Example 2

1 Investigation: Investigating the factors that affect how quickly household limescale removers work in removing limescale from an appliance

You may have to decide on which acids to use from a range that you are provided with. You would choose a weak acid, or weak acids, to test, rather than a strong acid, such as concentrated sulfuric acid. This is because of safety reasons, but also because the acid might damage the appliance you were trying to clean. You would then have to select a range of concentrations of your chosen weak acid to test.

2 Investigation: How speed affects the stopping distance of a trolley in the lab

The range of speeds you would choose would clearly depend on the speeds that you could produce in the lab.

> **Assessment tip**
>
> Again, it is often best to carry out a trial run or preliminary investigation, or carry out research, to determine the range to be investigated.

## Concentration

You might be trying to find out the best, or optimum, concentration of a disinfectant to prevent the growth of bacteria.

The 'best' concentration would be the lowest in a range that prevented the growth of the bacteria. Concentrations higher than this would be just wasting disinfectant.

If, in a preliminary test, no bacteria were killed by the concentration you used, you would have to increase it (or test another disinfectant). However, if there was no growth of bacteria in your preliminary test, you would have to lower the concentration range. A starting point might be to look at concentrations around those recommended by the manufacturer.

## ✹ Dependent variables

The dependent variable may be clear from the problem that you are investigating, for example the stopping distance of moving objects. You may have to make a choice.

> **Assessment tip**
>
> The value of the *depend*ent variable is likely to *depend* on the value of the independent variable. This is a good way of remembering the definition of a dependent variable.

### Example 3

1 Investigation: Measuring the rate of photosynthesis in a plant

There are several ways in which you could measure the rate of photosynthesis in a plant. These include:

> counting the number of bubbles of oxygen produced in a minute by a water plant such as *Elodea* or *Cabomba*

> measuring the volume of oxygen produced over several days by a water plant such as *Elodea* or *Cabomba*

> monitoring the concentration of oxygen in a polythene bag enclosing a potted plant, using a carbon dioxide sensor

> measuring the colour change of hydrogencarbonate indicator that contains algae embedded in gel.

## 2 Investigation: Measuring the rate of a chemical reaction

You could measure the rate of a chemical reaction in the following ways:

> the rate of formation of a product

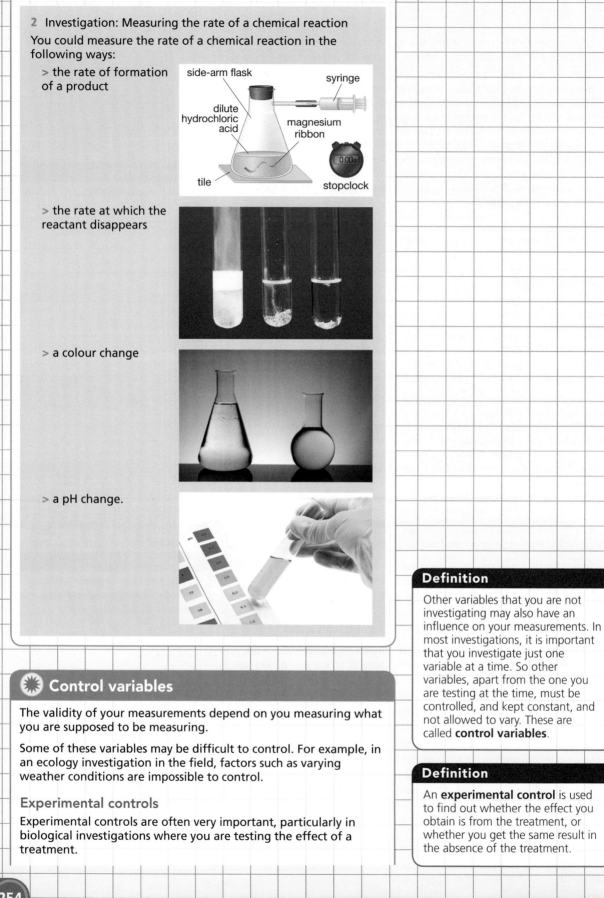

side-arm flask

syringe

dilute hydrochloric acid

magnesium ribbon

tile

stopclock

> the rate at which the reactant disappears

> a colour change

> a pH change.

### Control variables

The validity of your measurements depend on you measuring what you are supposed to be measuring.

Some of these variables may be difficult to control. For example, in an ecology investigation in the field, factors such as varying weather conditions are impossible to control.

#### Experimental controls

Experimental controls are often very important, particularly in biological investigations where you are testing the effect of a treatment.

**Definition**

Other variables that you are not investigating may also have an influence on your measurements. In most investigations, it is important that you investigate just one variable at a time. So other variables, apart from the one you are testing at the time, must be controlled, and kept constant, and not allowed to vary. These are called **control variables**.

**Definition**

An **experimental control** is used to find out whether the effect you obtain is from the treatment, or whether you get the same result in the absence of the treatment.

*Example 4*

Investigation: The effect of disinfectants on the growth of bacteria

If the bacteria do not grow, it could be because they have been killed by the disinfectant. The bacteria in your investigation may have died for some other reason. Another factor may be involved. To test whether any effects were down to the disinfectant, you need to set up the same practical, but this time using distilled water in place of the disinfectant. The distilled water is your control. If the bacteria are inhibited by the disinfectant, but grow normally in the dish containing distilled water, it is reasonable to assume that the disinfectant inhibited their growth.

# ✺ Assessing and managing risk

Before you begin any practical work, you must assess and minimise the possible risks involved.

Before you carry out an investigation, you must identify the possible hazards. These can be grouped into biological hazards, chemical hazards and physical hazards.

| **Biological hazards include:** | **Chemical hazards can be grouped into:** | **Physical hazards include:** |
|---|---|---|
| > microorganisms<br>> body fluids<br>> animals and plants. | > irritant and harmful<br>> toxic<br>> oxidising<br>> corrosive<br>> harmful to the environment. | > equipment<br>> objects<br>> radiation. |

Scientists use an international series of symbols so that investigators can identify hazards.

Hazards pose risks to the person carrying out the investigation.

A risk posed by concentrated sulfuric acid, for example, will be lower if you are adding one drop of it to a reaction mixture to make an ester, than if you are mixing a large volume of it with water.

When you use hazardous materials, chemicals or equipment in the laboratory, you must use them in such a way as to keep the risks to absolute minimum. For example, one way is to wear eye protection when using hydrochloric acid.

Any action that you carry out to reduce the risk of a hazard happening is known as a 'control measure'.

**Definition**

A **hazard** is something that has the potential to cause harm. Even substances, organisms and equipment that we think of being harmless, used in the wrong way, may be hazardous.

Hazard symbols are used on chemical bottles so that hazards can be identified.

**Definition**

The **risk** is the likelihood of a hazard causing harm in the circumstances it's being used in.

## ✳ Risk assessment

Before you begin an investigation, you must carry out a risk assessment. Your risk assessment must include:

✔ all relevant hazards (use the correct terms to describe each hazard, and make sure you include them all, even if you think they will pose minimal risk)

✔ risks associated with these hazards

✔ ways in which the risks can be minimised

✔ results of research into emergency procedures that you may have to take if something goes wrong.

You should also consider what to do at the end of the practical. For example, used agar plates should be left for a technician to sterilise; solutions of heavy metals should be collected in a bottle and disposed of safely.

### Assessment tip

To make sure that your risk assessment is full and appropriate:

> Remember that, for a risk assessment for a chemical reaction, the risk assessment should be carried out for the products and the reactants.

> When using chemicals, make sure the hazard and ways of minimising risk match the concentration of the chemical you are using; many acids, for instance, while being corrosive in higher concentrations, are harmful or irritant at low concentrations.

## ✳ Collecting primary data

✔ You should make sure that observations, if appropriate, are recorded in detail. For example, it is worth recording the appearance of your potato chips in your osmosis practical, in addition to the measurements you make.

✔ Measurements should be recorded in tables. Have one ready so that you can record your readings as you carry out the practical work.

✔ Think about the dependent variable and define this carefully in your column headings.

✔ You should make sure that the table headings describe properly the type of measurements that you have made, for example 'time taken for magnesium ribbon to dissolve'.

✔ It is also essential that you include units – your results are meaningless without these.

✔ The units should appear in the column head, and not be repeated in each row of the table.

### Definition

When you carry out an investigation, the data you collect are called **primary data.** The term 'data' is normally used to include your observations as well as measurements you might make.

### Definition

One set of results from your investigation may not reflect what truly happens. Carrying out repeats enables you to identify any results that don't fit. These are called **outliers** or **anomalous results**.

### Definition

If, when you carry out the same experiment several times, and get the same, or very similar results, the results are **repeatable**.

### Definition

Taking more than one set of results will improve the **reliability** of your data.

## ✳ Repeatability and reproducibility of results

When making measurements, in most instances, it is essential that you carry out repeats.

These repeats are one way of checking your results.

Results will not be repeatable of course, if you allow the conditions the investigation is carried out in to change.

You need to make sure that you carry out sufficient repeats, but not too many. In a titration, for example, if you obtain two values that are within $0.1\,cm^3$ of each other, carrying out any more will not improve the reliability of your results.

This is particularly important when scientists are carrying out scientific research and make new discoveries.

### Definition

The **reproducibility** of data is the ability of the results of an investigation to be reproduced by:

> using a different method and reaching the same conclusion

> someone else, who may be in a different lab, carrying out the same work.

## ✳ Processing data

### Calculating the mean

Using your repeat measurements you can calculate the arithmetical mean (or just 'mean') of these data. Often, the mean is called the 'average.'

| Temperature (°C) | Number of yeast cells (mm³) | | | Mean number of yeast cells (mm³) |
|---|---|---|---|---|
| | Test 1 | Test 2 | Test 3 | |
| 10 | 1000 | 1040 | 1200 | 1080 |
| 20 | 2400 | 2200 | 2300 | 2300 |
| 30 | 4600 | 5000 | 4800 | 4800 |
| 40 | 4800 | 5000 | 5200 | 5000 |
| 50 | 200 | 1200 | 700 | 700 |

You may also be required to use equations when processing data. Sometimes, these will need rearranging to be able to make the calculation you need. Practise using and rearranging equations as part of your preparation for assessment.

> **Definition**
>
> The **mean** is calculated by adding together all the measurements, and dividing by the number of measurements.

### Significant figures

When calculating the mean, you should be aware of significant figures.

For example, for the set of data below:

| 18 | 13 | 17 | 15 | 14 | 16 | 15 | 14 | 13 | 18 |
|---|---|---|---|---|---|---|---|---|---|

The total for the data set is 153, and ten measurements have been made. The mean is 15, and not 15.3.

This is because each of the recorded values has two significant figures. The answer must therefore have two significant figures. An answer cannot have more significant figures than the number being multiplied or divided.

> **Definition**
>
> **Significant figures** are the number of digits in a number based on the precision of your measurements.

### Using your data

When calculating means (and displaying data), you should be careful to look out for any data that do not fit in with the general pattern.

It might be the consequence of an error made in measurement. But sometimes outliers are genuine results. If you think an outlier has been introduced by careless practical work, you should ignore it when calculating the mean. But you should examine possible reasons carefully before just leaving it out.

> **Definition**
>
> An **outlier** (or **anomalous result**) is a reading that is very different from the rest.

## Displaying your data

Displaying your data – usually the means – makes it easy to pick out and show any patterns. It also helps you to pick out any anomalous data.

It is likely that you will have recorded your results in tables, and you could also use additional tables to summarise your results. The most usual way of displaying data is to use graphs. The table will help you decide which type to use.

| Type of graph | When you would use the graph | Example |
|---|---|---|
| bar chart or bar graph | where one of the variables is categoric | 'the diameters of the clear zones where the growth of bacteria was inhibited by different types of disinfectant' |
| line graph | where independent and dependent variables are both continuous | 'the volume of carbon dioxide produced by a range of different concentrations of hydrochloric acid' |
| scatter graph | to show an association between two (or more) variables | 'the association between length and breadth of a number of privet leaves'<br><br>In scatter graphs, the points are plotted, but not usually joined. |

If it is possible from the data, join the points of a line graph using a straight line, or in some instances, a curve. In this way, graphs can also help you to process data.

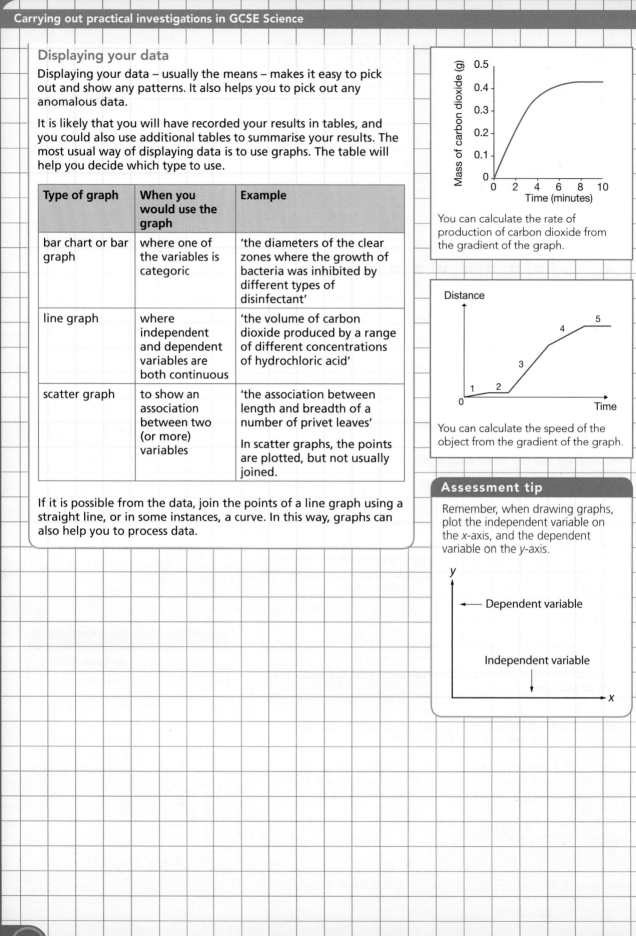

You can calculate the rate of production of carbon dioxide from the gradient of the graph.

You can calculate the speed of the object from the gradient of the graph.

### Assessment tip

Remember, when drawing graphs, plot the independent variable on the x-axis, and the dependent variable on the y-axis.

# ✳ Conclusions from differences in data sets

When comparing two (or more) sets of data, you can often compare the values of two sets of means.

## Example 5

Investigation: Comparing the effectiveness of two disinfectants

Two groups of students compared the effectiveness of two disinfectants, labelled A and B. Their results are shown in the table.

| Disinfectant | Diameter of zone of inhibition (clear zone) (mm) | | | | | | | | | | Mean dia (mm) |
|---|---|---|---|---|---|---|---|---|---|---|---|
| | 1 | 2 | 3 | 4 | 5 | 6 | 7 | 8 | 9 | 10 | |
| A | 15 | 13 | 17 | 15 | 14 | 16 | 15 | 14 | 13 | 18 | 15 |
| B | 25 | 23 | 24 | 23 | 26 | 27 | 25 | 24 | 23 | 22 | 24 |

When the means are compared, it appears that disinfectant B is more effective in inhibiting the growth of bacteria. Can you be sure? The differences might have resulted from the treatment of the bacteria using the two disinfectants. But the differences could have occurred purely by chance.

Scientists use statistics to find the probability of any differences having occurred by chance. The lower this probability is, which is found out by statistical calculations, the more likely it is that it was (in this case) the disinfectant that caused the differences observed.

Statistical analysis can help to increase the confidence you have in your conclusions.

> **Assessment tip**
>
> You have learned about probability in maths lessons.

> **Definition**
>
> If there is a relationship between dependent and independent variables that can be defined, there is a **correlation** between the variables.

# ✳ Drawing conclusions

Observing trends in data or graphs will help you to draw conclusions. You may obtain a linear relationship between two sets of variables, or the relationship might be more complex.

## Example 6

**Conclusion:** The higher the concentration of acid, the shorter the time taken for the magnesium ribbon to dissolve.

**Conclusion:** The higher the concentration of acid, the faster the rate of reaction.

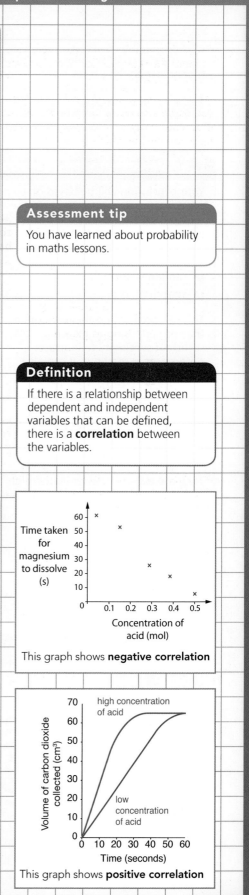

This graph shows **negative correlation**

This graph shows **positive correlation**

When drawing conclusions, you should try to relate your findings to the science involved.

> In the first investigation in Example 6, your discussion should focus on the greater possibility/increased frequency of collisions between reacting particles as the concentration of the acid is increased.

> In the second investigation in Example 6, there is a clear scientific mechanism to link the rate of reaction to the concentration of acid.

Sometimes, you can see correlations between data which are coincidental, where the independent variable is not the cause of the trend in the data.

*Example 7*

Studies have shown that levels of vitamin D are very low in people with long-term inflammatory diseases. But there's no scientific evidence to suggest that these low levels are the cause of the diseases.

## ✳ Evaluating your investigation

Your conclusion will be based on your findings, but must take into consideration any uncertainty in these introduced by any possible sources of error. You should discuss where these have come from in your evaluation.

The two types of errors are:

✔ random error
✔ systematic error.

This can occur when the instrument that you are using to measure lacks sufficient sensitivity to indicate differences in readings. It can also occur when it is difficult to make a measurement. If two investigators measure the height of a plant, for example, they might choose different points on the compost, and the tip of the growing point to make their measurements.

They are either consistently too high or too low. One reason could be down to the way you are making a reading, for example taking a burette reading at the wrong point on the meniscus. Another could be the result of an instrument being incorrectly calibrated, or not being calibrated.

The volume of liquid in a burette must be read to the bottom of the meniscus

### Definition

**Error** is a difference between a measurement you make, and its true value.

### Definition

With **random error**, measurements vary in an unpredictable way.

### Definition

With **systematic error**, readings vary in a controlled way.

### Assessment tip

A pH meter must be calibrated before use using buffers of known pH.

### Assessment tip

What you should not discuss in your evaluation are problems introduced by using faulty equipment, or by you using the equipment inappropriately. These errors can, or could have been, eliminated, by:

> checking equipment

> practising techniques before your investigation

> taking care and patience when carrying out the practical.

## ✳ Accuracy and precision

When evaluating your investigation, you should mention accuracy and precision. If you use these terms, it is important that you understand what they mean, and that you use them correctly.

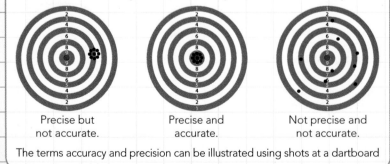

| Precise but not accurate. | Precise and accurate. | Not precise and not accurate. |

The terms accuracy and precision can be illustrated using shots at a dartboard

### Definition

When making measurements:
> the **accuracy** of the measurement is how close it is to the true value

> **precision** is how closely a series of measurements agree with each other.

## ✹ Improving your investigation

When evaluating your investigation, you should discuss how your investigation could be improved. This could be by improving:

✔ the reliability of your data. For example, you could make more repeats, or more frequent readings, or 'fine-tune' the range you chose to investigate, or refine your technique in some other way.

✔ the accuracy and precision of your data, by using measuring equipment with a higher resolution.

In science, the measurements that you make as part of your investigation should be as precise as you can, or need to, make them. To achieve this, you should use:

✔ the most appropriate measuring instrument

✔ the measuring instrument with the most appropriate size of divisions.

The smaller the divisions you work with, the more precise your measurements. For example:

✔ In an investigation on how your heart rate is affected by exercise, you might decide to investigate this after a 100 m run. You might measure out the 100 m distance using a trundle wheel, which is sufficiently precise for your investigation.

✔ In an investigation on how light intensity is affected by distance, you would make your measurements of distance using a metre rule with millimetre divisions; clearly a trundle wheel would be too imprecise.

✔ In an investigation on plant growth, in which you measure the thickness of a plant stem, you would use a micrometer or Vernier callipers. In this instance, a metre rule would be too imprecise.

## ✹ Using secondary data

As part of controlled assessment, you will be expected to compare your data – primary data – with **secondary data** you have collected.

One of the simplest ways of doing this is to compare your data with other groups in your class who have carried out an identical practical investigation.

In your controlled assessment, you will be provided with a data sheet of secondary data.

You should also, if possible, search through the scientific literature – in textbooks, the internet, and databases – to find data from similar or identical practical investigations so that you can compare the data with yours.

Ideally, you should use secondary data from a number of sources, carry out a full analysis of the secondary data that you have collected, and compare the findings with yours.

You should critically analyse any evidence that conflicts with yours, and suggest and discuss what further data might help to make your conclusions more secure.

### Definition

**Secondary data** are measurements/observations made by anyone other than you.

You should review secondary data and evaluate it. Scientific studies are sometimes influenced by the **bias** of the experimenter.

✔ One kind of bias is having a strong opinion related to the investigation, and perhaps selecting only the results that fit with a hypothesis or prediction.

✔ The bias could be unintentional. In fields of science that are not yet fully understood, experimenters may try to fit their findings to current knowledge and thinking.

There have been other instances where the 'findings' of experimenters have been influenced by organisations that supplied the funding for the research.

You must fully reference any secondary data that you have used, using one of the accepted referencing methods.

## ✳ Referencing methods

The two main conventions for writing a reference are the:

✔ Harvard system
✔ Vancouver system.

In your text, the Harvard system refers to the authors of the reference, for example 'Smith and Jones (1978)'.

The Vancouver system refers to the number of the numbered reference in your text, for example '... the reason for this hypothesis is unknown.[5]'.

Though the Harvard system is usually preferred by scientists, it is more straightforward for you to use the Vancouver system.

### Harvard system

In your references list a book reference should be written:

Author(s) (year of publication). *Title of Book*, publisher, publisher location.

The references are listed in alphabetical order according to the authors.

### Vancouver system

In your references list a book reference should be written:

1 Author(s). *Title of Book*. Publisher, publisher location: year of publication.

The references are numbered in the order in which they are cited in the text.

### Assessment tip

Remember to write out the URL of a website in full. You should also quote the date when you looked at the website.

## ❋ Do the data support your hypothesis?

You need to discuss, in detail, whether all, or which of your primary, and the secondary data you have collected, support your original hypothesis. They may, or may not.

You should communicate your points clearly, using the appropriate scientific terms, and checking carefully your use of spelling, punctuation and grammar. You will be assessed on this written communication as well as your science.

If your data do not completely match your hypothesis, it may be possible to modify the hypothesis or suggest an alternative one. You should suggest any further investigations that can be carried out to support your original hypothesis or the modified version.

It is important to remember, however, that if your investigation does support your hypothesis, it can improve the confidence you have in your conclusions and scientific explanations, but it can't prove your explanations are correct.

## ❋ Your controlled assessment

The assessment of your investigation will form part of what is called controlled assessment. AQA will provide the task for you to investigate.

You may be able to work in small groups to carry out the practical work, but you will have to work on your own to write up your investigation.

The Controlled Assessment is worth 25% of the marks for your GCSE Science. It is worth doing it well!

You will probably have between seven and eight lessons (approximately eight hours) to complete your Controlled Assessment:

> one to two for the planning stage

> one to two for the practical work

> three to four for the analysis stage.

You will be expected to plan your investigation independently and to make and record observations with precision and accuracy. As well as your own data you will need to analyse some obtained by other people and compare them with yours.

You will need to write a final report, along with your plan and risk assessment. It needs to provide evidence that you can:

✔ research and plan

✔ assess and manage risks

✔ collect, process and analyse data

✔ evaluate your practical activity.

# How to be successful in your GCSE Science assessment

## Introduction

*AQA uses assessments to test how good your understanding of scientific ideas is, how well you can apply your understanding to new situations and how well you can analyse and interpret information you have been given. The assessments are opportunities to show how well you can do these.*

To be successful in exams you need to:

✔ have a good knowledge and understanding of science
✔ be able to apply this knowledge and understanding to familiar and new
✔ situations
✔ be able to interpret and evaluate evidence that you have just been given.

You need to be able to do these things under exam conditions.

### ✴ The language of the assessment paper

When working through an assessment paper, make sure that you:

✔ re-read a question enough times until you understand exactly what the examiner is looking for
✔ make sure that you highlight key words in a question. In some instances, you will be given key words to include in your answer.
✔ look at how many marks are allocated for each part of a question. In general, you need to write at least as many separate points in your answer as there are marks.

### ✴ What verbs are used in the question?

A good technique is to see which verbs are used in the wording of the question and to use these to gauge the type of response you need to give. The table lists some of the common verbs found in questions, the types of responses expected and then gives an example.

| Verb used in question | Response expected in answer | Example question |
|---|---|---|
| write down<br><br>state<br><br>give<br><br>identify | These are usually more straightforward types of question in which you are asked to give a definition, make a list of examples, or the best answer from a series of options. | 'Write down three types of microorganism that cause disease.'<br><br>'State one difference and one similarity between radio waves and gamma rays.' |
| calculate | Use maths to solve a numerical problem. | 'Calculate the cost of supplying the flu vaccine to the whole population of the UK.' |

| estimate | use maths to solve a numerical problem, but you do not have to work out the exact answer | 'Estimate the number of bacteria in the culture after five hours.' |
|---|---|---|
| describe | use words (or diagrams) to show the characteristics, properties or features of, or build an image of something | 'Describe how antibiotic resistance can be reduced.' |
| suggest | come up with an idea to explain information that you are given | 'Suggest why eating fast foods, rather than wholegrain foods, could increase the risk of obesity.' |
| demonstrate show how | use words to make something evident using reasoning | 'Show how enzyme activity changes with temperature.' |
| compare | look for similarities and differences | 'Compare the structure of arteries and veins.' |
| explain | to offer a reason for, or make understandable, information that you are given | 'Explain why measles cannot be treated with antibiotics.' |
| evaluate | to examine and make a judgement about an investigation or information that you are given | 'Evaluate the evidence for vaccines causing harm to human health.' |

## ✹ What is the style of the question?

Try to get used to answering questions that have been written in lots of different styles before you sit the exam. Work through past papers, or specimen papers, to get a feel for these. The types of questions in your assessment fit the three assessment objectives shown in the table.

| Assessment objective |
|---|
| **AO1** Recall, select and communicate your knowledge and understanding of science. |
| **AO2** Apply skills, knowledge and understanding of science in practical and other contexts. |
| **AO3** Analyse and evaluate evidence, make reasoned judgements and draw conclusions based on evidence. |

### Assessment tip

Of course you must revise the subject material adequately. It is as important that you are familiar with the different question styles used in the exam paper, as well as the question content.

## How to answer questions on: AO1 Recall the science

These questions, or parts of questions, test your ability to recall your knowledge of a topic. There are several types of this style of question:

✔ Fill in the spaces (you may be given words to choose from)
✔ Tick the correct statements
✔ Use lines to link a term with its definition or correct statement
✔ Add labels to a diagram
✔ Complete a table
✔ Describe a process

*Example 8*

a What is meant by the term *metabolic rate*?
  Tick (✓) **one** box.
  ☐ the amount of energy a person uses each hour
  ☐ the amount of exercise a person does each day
  ☐ the amount of food a person eats each day

## How to answer questions on: AO1 Recall the science in practical techniques

You may be asked to recall how to carry out certain practical techniques, either ones that you have carried out before, or techniques that scientists use.

To revise for these types of questions, make sure that you have learned definitions and scientific terms. Produce a glossary of these, or key facts cards, to make them easier to remember. Make sure that your key facts cards also cover important practical techniques, including equipment, where appropriate.

*Example 9*

Describe how to test the pH of a solution.

**Assessment tip**

Do not forget that mind maps – either drawn by you or by using a computer program – are very helpful when revising key points.

## How to answer questions on: AO2 Apply skills, knowledge and understanding

Some questions require you to apply basic knowledge and understanding in your answers.

You may be presented with a topic that is familiar to you, but you should also expect questions in your Science exam to be set in an unfamiliar context.

Questions may be presented as:

✔ practical investigations
✔ data for you to interpret
✔ a short paragraph or article.

The information required for you to answer the question might be in the question itself, but for later stages of the question, you may be asked to draw on your knowledge and understanding of the subject material in the question.

Practice will help you to become familiar with contexts that examiners use and question styles. However, you will not be able to predict many of the contexts used. This is deliberate; being able to apply your knowledge and understanding to different and unfamiliar situations is a skill the examiner tests.

Practise doing questions where you are tested on being able to apply your scientific knowledge and your ability to understand new situations that may not be familiar. In this way, when this type of question comes up in your exam, you will be able to tackle it successfully.

### Assessment tip

Work through the Preparing for assessment: Applying your knowledge tasks in this book as practice.

### Example 10

Measles is an infectious disease caused by a virus. Today, most children are vaccinated against measles when they are very young.

The graph shows the number of measles cases per year in the USA between 1950 and 2000. The graph also shows when vaccination against measles was first introduced in the USA.

The use of the measles vaccine reduces the number of measles cases.

Explain why this graph alone does not prove that the use of the measles vaccine reduces the number of measles cases.

## ✹ How to answer questions on: AO2 Apply skills, knowledge and understanding in practical investigations

Some opportunities to demonstrate your application of skills, knowledge and understanding will be based on practical investigations. You may have carried out some of these investigations, but others will be new to you, and based on data obtained by scientists. You will be expected to describe patterns in data from graphs you are given or that you will have to draw from given data.

Again, you will have to apply your scientific knowledge and understanding to answer the question.

### Example 11

Look at the graph on the right showing the resistance of the bacterium, *Streptococcus pneumoniae*, to three different types of antibiotic.

a Which antibiotic does there seem to be least resistance to, even when it has been used before?

b Can you explain why this might be the case?

You will also need to analyse scientific evidence or data given to you in the question. It is likely that you will not be familiar with the material.

Analysing data may involve drawing graphs and interpreting them, and carrying out calculations. Practise drawing and interpreting graphs from data.

When drawing a graph, make sure that you:

✔ choose and label the axes fully and correctly

✔ include units, if this has not been done already

✔ plot points on the graph carefully – the examiner will check individual points to make sure that they are accurate

✔ join the points correctly; usually this will be by a line of best fit.

When reading values off a graph that you have drawn or one given in the question, make sure that you:

✔ do it carefully, reading the values as accurately as you can

✔ double-check the values.

When describing patterns and trends in the data, make sure that you:

✔ write about a pattern or trend in as much detail as you can

✔ mention anomalies where appropriate

✔ recognise there may be one general trend in the graph, where the variables show positive or negative correlation

✔ recognise the data may show a more complex relationship. The graph may demonstrate different trends in several sections. You should describe what's happening in each.

✔ describe fully what the data show.

## ✸ How to answer questions needing calculations

✔ The calculations that you are asked to do may be straightforward – for example the calculation of the mean from a set of data.

✔ They may be more complex – for example calculating the yield of a chemical reaction.

✔ Other questions will require the use of formulae.

You will be given an equation sheet with the question paper.

On page 271, there is a list of the maths skills that you will need. Remember, these are the same skills that you have learned in maths lessons.

### Example 12

Calculate the area on the agar plate, around the antibiotic disc, that is free from bacteria.

Use the formula:

area = $\pi r^2$

where $\pi$ = 3.14

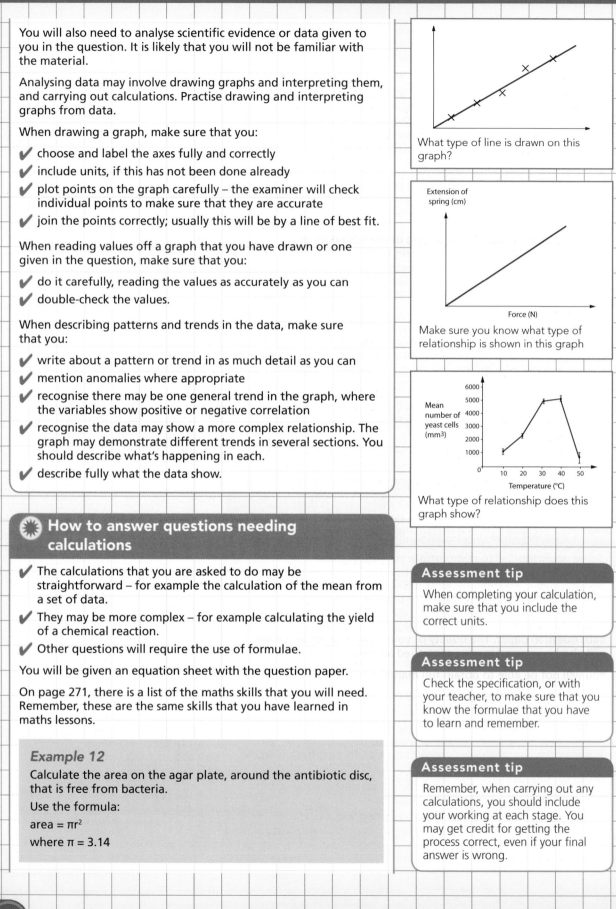

What type of line is drawn on this graph?

Extension of spring (cm)

Force (N)

Make sure you know what type of relationship is shown in this graph

Mean number of yeast cells (mm3)

Temperature (°C)

What type of relationship does this graph show?

**Assessment tip**

When completing your calculation, make sure that you include the correct units.

**Assessment tip**

Check the specification, or with your teacher, to make sure that you know the formulae that you have to learn and remember.

**Assessment tip**

Remember, when carrying out any calculations, you should include your working at each stage. You may get credit for getting the process correct, even if your final answer is wrong.

# How to answer questions on: AO3 Analysing and evaluating evidence

For these types of questions, in addition to analysing data, you must also be able to evaluate information that you are given. This is one of the hardest skills. Think about the validity of the scientific data: did the technique(s) used in any practical investigation allow the collection of accurate and precise data?

Your critical evaluation of scientific data in class, along with the practical work and controlled assessment work, will help you to develop the evaluation skills required for these types of questions.

### Example 13

In the standard experiment testing of the inhibition of bacterial growth by a new antibiotic, explain why further investigation is required to confirm the effectiveness of the antibiotic.

You may be expected to compare data with other data, or come to a conclusion about its reliability, its usefulness or its implications. Again, it is possible that you will not be familiar with the context. You may be asked to make a judgement about the evidence or to give an opinion with reasons.

### Example 14

A catalytic converter reduces nitrogen oxide emissions from a car engine to nitrogen. Evidence shows that the reaction needs the catalyst to be hot. Explain the effect of this on air quality in housing estates where many people use their car to commute to work, and suggest the possible implications for health.

**Assessment tip**

Work through the Preparing for assessment: Analysing data tasks, in this book, as practice.

**Assessment tip**

Wherever possible, use as much data as you can in your answer, particularly when explaining trends or conclusions, so you can gain full marks. Try to use numbers and values rather than just trends in data or graphs. 'At 45 °C…' is always better than 'as the temperature rises it gets greater'.

# The quality of your written communication

Scientists need good communication skills to present and discuss their findings. You will be expected to demonstrate these skills in the exam. You will be assessed in the longer-response exam questions that you answer. These questions are clearly indicated in each question paper. The quality of your written communication will also be assessed in your controlled assessment.

You will not be able to obtain full marks unless you:

✔ make sure that the text you write is legible

✔ make sure that spelling, punctuation and grammar are accurate so that the meaning of what you write is clear

✔ use a form and style of writing appropriate for its purpose and for the complexity of the subject matter

✔ organise information clearly and coherently

✔ use the scientific language correctly.

You will also need to remember the writing and communication skills that you have developed in English lessons. For example, make sure that you understand how to construct a good sentence using connectives.

**Assessment tip**

You will be assessed on the way in which you communicate science ideas.

**Assessment tip**

When answering questions, you must make sure that your writing is legible. An examiner cannot award marks for answers that he or she cannot read.

## ✳ Revising for your Science exam

You should revise in the way that suits you best. It is important that you plan your revision carefully, and it is best to start well before the date of the exams. Take the time to prepare a revision timetable and try to stick to it. Use this during the lead up to the exams and between each exam.

When revising:

✔ Find a quiet and comfortable space in the house where you will not be disturbed. It is best if it is well ventilated and has plenty of light.

✔ Take regular breaks. Some evidence suggests that revision is most effective when you revise in 30 to 40 minute slots. If you get bogged down at any point, take a break and go back to it later when you're feeling fresh. Try not to revise when you are feeling tired. If you do feel tired, take a break.

✔ Use your school notes, textbook and possibly a revision guide. But also make sure that you spend some time using past papers to familiarise yourself with the exam format.

✔ Produce summaries of each topic.

✔ Draw mind maps covering the key information on a topic.

✔ Set up revision cards containing condensed versions of your notes.

✔ Ask yourself questions, and try to predict questions, as you are revising a topic.

✔ Test yourself as you go along. Try to draw key labelled diagrams, and try some questions under timed conditions.

✔ Prioritise your revision of topics. You might want to allocate more time to revising the topics you find most difficult.

> **Assessment tip**
>
> Try to make your revision timetable as specific as possible – don't just say 'science on Monday, and Thursday', but list the topics that you will cover on those days.

> **Assessment tip**
>
> Start your revision well before the date of the exams, produce a revision timetable, and use the revision strategies that suit your style of learning. Above all, revision should be an active process.

## ✳ How do I use my time effectively in the exam?

Timing is important when you sit an exam. Do not spend so long on some questions that you leave insufficient time to answer others. For example, in a 60-mark question paper, lasting one hour, you will have, on average, one minute per question.

If you are unsure about certain questions, complete the ones you are able to do first, then go back to the ones you're less sure of.

If you have time, go back and check your answers at the end of the exam.

## ✳ On exam day...

A little bit of nervousness before your exam can be a good thing, but try not to let it affect your performance in the exam. When you turn over the exam paper keep calm. Look at the paper and get it clear in your head exactly what is required from each question. Read each question carefully. Do not rush.

If you read a question and think that you have not covered the topic, keep calm – it could be that the information needed to answer the question is in the question itself or the examiner may be asking you to apply your knowledge to a new situation.

Finally, good luck!

## ✳ Mathematical skills

You will be allowed to use a calculator in all assessments.

These are the maths skills that you need, to complete all the assessments successfully.

You should understand:

✔ the relationship between units, for example, between a gram, kilogram and tonne

✔ compound measures such as speed

✔ when and how to use estimation

✔ the symbols = < > ~

✔ direct proportion and simple ratios

✔ the idea of probability.

You should be able to:

✔ give answers to an appropriate number of significant figures

✔ substitute values into formulae and equations using appropriate units

✔ select suitable scales for the axes of graphs

✔ plot and draw line graphs, bar charts, pie charts, scatter graphs and histograms

✔ extract and interpret information from charts, graphs and tables.

You should be able to calculate:

✔ using decimals, fractions, percentages and number powers, such as $10^3$

✔ arithmetic means

✔ areas, perimeters and volumes of simple shapes

**In addition, if you are a higher tier candidate, you should be able to:**

✔ **change the subject of an equation**

**and should be able to use:**

✔ **numbers written in standard form**

✔ **calculations involving negative powers, such as $10^{-1}$**

✔ **inverse proportion**

✔ **percentiles and deciles.**

## ✳ Some key equations

With the written paper for Unit 2, there will be an equation sheet. In order to make best use of the sheet, it will help if you practise using the following equations.

| Use of equation | Equation and its units |
|---|---|
| power | 1 watt = 1 joule/second |
| power consumed by an electrical appliance | power = potential difference × current<br>watts          volts          amps<br>W          V          A |
| for an electrical appliance in the home | power = energy transferred ÷ time<br>kilowatt          kilowatt-hour          hours<br>kW          kWh          h |
| | power = energy transferred ÷ time<br>watts          joules          seconds<br>W          J          s |
| in a domestic electricity meter | 1 unit of electricity = 1 kWh |
| cost of using an electrical appliance | total cost = number of kilowatt-hours × cost per kilowatt-hour |
| efficiency of an appliance | $\text{efficiency} = \dfrac{\text{useful energy out}}{\text{total energy in}} \times (100\%)$ |
| | $\text{efficiency} = \dfrac{\text{useful power out}}{\text{total power in}} \times (100\%)$ |
| electromagnetic waves | velocity = frequency × wavelength<br>metres per second          hertz          metres<br>m/s          Hz          m |

# Useful data for GCSE Science

## Introduction

*In all your GCSE Science assessments, you will need to know and apply units and symbols correctly. You will also need to recognise and use the names and formulae of some elements and compounds.*

*This will help to show the examiners how well you understand the science that you are using.*

### ✸ Units, symbols and chemical compounds

You need to be able to use the correct units and symbols when measuring and calculating.

| Quantity | Unit | Symbol |
|---|---|---|
| mass | kilogram | kg |
| | gram | g |
| | milligram | mg |
| | microgram | µg |
| length | kilometre | km |
| | metre | m |
| | centimetre | cm |
| | millimetre | mm |
| | micrometre | µm |
| volume | cubic metre | $m^3$ |
| | cubic decimetre | $dm^3$ |
| | litre | ℓ |
| | cubic centimetre | $cm^3$ |
| | millilitre | mℓ |
| time | hour | h |
| | minute | min |
| | second | s |
| temperature | degrees Celsius | °C |
| chemical quantity | mole | mol |
| potential difference (voltage) | volt | V |
| current | ampere | A |
| | milliampere | mA |
| force | newton | N |
| energy/work | kilojoule | kJ |
| | joule | J |
| | kilowatt-hour | kWh |

| Quantity | Unit | Symbol |
|---|---|---|
| power | kilowatt | kW |
| | watt | W |
| frequency | hertz | Hz |
| wavelength | metre | m |
| velocity (wave speed) | metre per second | m/s |

You need to know the chemical symbols for the first 20 elements in the periodic table.

You need to know the chemical symbols of the following metals and non-metals.

| Element | Symbol |
|---|---|
| **Metals** | |
| copper | Cu |
| gold | Au |
| iron | Fe |
| lead | Pb |
| nickel | Ni |
| silver | Ag |
| zinc | Zn |
| **Non-metals** | |
| sulfur | S |

You need to know the names and formulae of the following chemical compounds.

| Compound | Formula |
|---|---|
| aluminium hydroxide | $Al(OH)_3$ |
| ammonia | $NH_3$ |
| calcium carbonate | $CaCO_3$ |
| calcium hydroxide | $Ca(OH)_2$ |
| calcium oxide | $CaO$ |
| carbon dioxide | $CO_2$ |
| chlorine gas | $Cl_2$ |
| hydrogen gas | $H_2$ |
| hydrogen chloride | $HCl$ |
| magnesium hydroxide | $Mg(OH)_2$ |
| methane | $CH_4$ |
| nitrogen gas | $N_2$ |
| oxygen gas | $O_2$ |
| sodium bicarbonate | $NaHCO_3$ |
| sodium chloride | $NaCl$ |
| water | $H_2O$ |

## Bad Science for Schools

### When the evidence doesn't add up.

Sometimes people use what sound like scientific words and ideas to sell you things or persuade you to think in a certain way. Some of these claims are valid, and some are not. The activities on these pages are based on the work of Dr Ben Goldacre and will help you to question some of the scientific claims you meet. Read more about the work of Ben in his *Bad Science* book or at badscience.net.

## How much to look younger?

There are many ways to make yourself look younger if you're an adult. These include the style and colour of your hair, the texture of your skin, your body shape and the clothes you wear. Manufacturers and retailers know this and recognise where there's money to be made.

Which of these do you think is more effective?

Are there other ways for adults to make themselves look younger?

How are these age-defying products promoted?

### ✳ YOUNG SKIN FROM OLD?

Skin changes in appearance as people get older. These photographs show how older skin looks different to young skin.

> Examine the photographs. What are the differences?

> How might an anti-ageing skin cream work on the old skin? What would it need to do?

 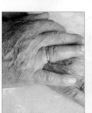

Young skin    Old skin

### ✳ THE SCIENCE BEHIND THE CLAIM

As you get older you may not like the appearance of wrinkles and crows feet. You can spend quite a lot of money on anti-ageing skin creams. Creams are advertised with appealing images and lavish claims, but do they really work?

One immediate gain from a cream is rehydration. Dried out skin doesn't look good so we can make it look better by moisturising it. This is easy and the active ingredients are really cheap. However something more is needed to make someone genuinely look younger.

These are three types of active ingredient commonly used:

> Alphahydroxy acids, such as vitamins A and C, are used to exfoliate the skin. Some of these work at high doses, but they are also irritants, so they can only be sold at low doses.

> Vegetable proteins, which are long chain molecules. As the cream dries on the skin the chain molecules tighten, applying tension and temporarily tightening it.

> Hydrogen peroxide, which is corrosive and will lightly burn the skin.

Why might someone who uses a cream with these ingredients think that it is working? Will the effects last?

# NEW AND IMPROVED! ADVERTISER'S CLAIMS

Many anti-ageing skin creams are sold on the basis that there is a scientific reason that they work. Some claims are justified but others are pretty dubious, even if they look persuasive at first glance. You should think critically about what you are told by the advertisers.

> Claims sound more convincing if they are based on tests and if scientists have been involved. Powerful scientific words include *conclusive tests*, *laboratory*, *cleanse*, *purify* and *health*.

> They may claim to make you feel better, look younger, have more energy and be healthier. Some of the claims may be difficult to prove; they should have been tested, the full results published and independently checked in a scientific way. There are very few cases of anti-ageing skin creams being proved to get rid of wrinkles. Why do you think this is?

> Watch out for claims such as 'eight out of ten users said that...' if it's not clear what kind of people and how many were asked. What do you think ten company employees might say about their product and would this be representative of all their consumers?

## ✳ THE PSYCHOLOGY OF COSMETICS

You can buy very cheap creams in the shops. You can even make your own skin cream using simple ingredients. If you did, it would be pretty good at moisturising, so your skin would feel soft and maybe a little smoother. It wouldn't, however, make you look younger for long.

Why do you think anti-ageing skin creams are sometimes quite expensive?

Do you think people who buy anti-ageing skin creams

a) genuinely believe that they make them look younger?

b) hope that they might but don't really believe it?

c) do it because it makes them feel good?

## Bad Science for Schools

### When the evidence doesn't add up.

Sometimes people use what sound like scientific words and ideas to sell you things or persuade you to think in a certain way. Some of these claims are valid, and some are not. The activities on these pages are based on the work of Dr Ben Goldacre and will help you to question some of the scientific claims you meet. Read more about the work of Ben in his *Bad Science* book or at badscience.net.

## Keeping your brain fit

In science you learn about ideas that scientists have developed by collecting evidence from experiments; you are also learning to collect and evaluate evidence yourself. You can use this outside of the laboratory to weigh up information you come across everyday. Let's look at this example about how to prepare for exams.

When it comes to exam season you will get lots of different tips from teachers, other students and companies that all claim to help you do better in exams but who is right? If you had an important exam coming up, what would be a good way of making sure that your brain was going to function well?

### ✳ GOOD ADVICE?

Here are three pieces of advice offered to students before exams. For each one:

> suggest why it might be true;

> suggest why you might be dubious about it;

> decide whether you think it's good advice and explain why.

Before sitting an aural exam (a listening test) spin round three times clockwise and three times anticlockwise. This stimulates the semicircular canals which are located in the inner ear, thus stimulating the cochlea.

A drink with caffeine in it is a good idea as it acts as a stimulant and will cause your brain to work quicker.

Before doing an exam in the morning make sure you have a good breakfast. Something like porridge is good as the energy is released slowly during the morning, so you don't get tired towards the end.

## ✱ THE SCIENCE BEHIND THE CLAIM

Let's look in more detail at some activities that some schools have used to try to improve students' concentration and learning. Your task is to work out which bits of science are good science and which are bad. To help you decide whether you think your activity is good or bad try discussing these questions:

> What *advice* is being given?

> What *claim* is being made?

> What *scientific ideas* are being used to justify that claim?

> What *scientific ideas* do you have that may tell you something about this topic?

> Is the advice *sound*?

This is the advice.

This is the claim.

Interlock the fingers of both your hands, holding your elbows out at the sides. This completes a circuit and allows positive energy to flow. Positive energy creates positive thoughts, stimulating the brain, stilling anxieties and clearing the way for a free flow of logical thought.

| These are the scientific ideas used to back up this claim. Forming the arms in a loop creates no circuit that any kind of energy 'flows around' and 'positive energy' is a meaningless term. | So can you think of any reason why this might work? You know that regular exercise is good and could help to refocus on ideas and mental activities. | Is the advice sound? Well, it won't do you any harm and may even improve concentration, but not for the reasons claimed. |

### Now you have a go. Are these good or bad science?

Water is a vital ingredient of blood and blood is essential to transport oxygen to the brain. For the brain to work well you have to ensure your blood is hydrated. This needs water, little and often. The best way of rehydrating the blood taking oxygen to the brain is to hold water in the mouth for up to half a minute, thus allowing direct absorption.

Your carotid arteries are vital to supplying your brain with richly oxygenated blood. Ensure their peak performance by pressing your brain buttons. These are just below the collar bone, one on either side. Make 'C' shapes with forefinger and thumb to place over the brain buttons and gently massage.

## ✱ WOULD YOU PAY MONEY FOR THIS?

Many products or services are sold on the basis that there is a scientific reason that they work. Some claims are justified but some are pretty dubious, even if they look persuasive at first glance. You should think critically about what you are told.

> Claims might sound more convincing if they use technical scientific terms, but sometimes they are used incorrectly, to make something sound scientific when it's not.

> Powerful scientific words include 'conclusive tests', 'energy', 'cleanse', 'purify' and 'health'.

> They may claim to make you feel better, look better, have more energy and be healthier.

> The claims may be true but they should have been tested, the full results published and independently checked.

# Glossary

**absorb** an object absorbs energy when the energy from infrared radiation is transferred to the particles of the object, increasing the temperature of the object

**adaptation** the way in which an organism evolved to become better able to survive in its environment

**addiction** when a person becomes dependent on a drug

**aerobic respiration** a process in which energy is released from glucose, using oxygen

**alcohols** a family of organic compounds containing an OH group, for example ethanol ($C_2H_5OH$)

**algae** simple, plant-like organisms

**alkanes** a family of hydrocarbons: $C_nH_{2n+2}$ with single covalent bonds – found in crude oil

**alkenes** a family of hydrocarbons: $C_nH_{2n}$ with double covalent bonds (C=C), for example ethene ($C_2H_4$)

**alleles** a particular form of a gene

**alloy** mixture of two or more metals, with useful properties different from the individual metals

**amino acids** small molecules from which proteins are built

**ampere (amp)** unit used to measure electrical current

**amplitude** size of wave oscillation – for a mechanical wave, it describes how far the particles vibrate around their central position

**anneal** use heating to remove internal stress, crystal defects and dislocations in a material

**anode** positive electrode

**antibiotic** therapeutic drug acting to kill bacteria which is taken into the body

**antibody** protein normally present in the body or produced in response to an antigen which it neutralises, thus producing an immune response

**antiseptic** substance that kills pathogens

**antitoxin** substance produced by white blood cells that neutralises the effects of toxins

**appliance** device that transfers the energy supplied by electricity into something useful

**argon** the most common noble gas – makes up nearly 1% of the air

**atmosphere** thin layer of gas surrounding a planet

**atom** the basic 'building block' of an element that cannot be chemically broken down

**atomic number** number of protons found in the nucleus of an atom

**attraction** force that pulls, or holds, objects together

**bacteria** single-celled microorganisms that can either be free-living organisms or parasites (they sometimes invade the body and cause disease)

**balanced equation** chemical equation where the number of atoms of each element is the same in the reactants as in the products

**Big Bang theory** theory that states that the universe originated from a point at very high temperature, and everything in the universe formed as energy and matter exploded outwards from that point and cooled down

**biodegradable** a biodegradable material can be broken down by microorganisms

**biodiversity** range of different living organisms in a habitat

**biomass** the mass of living material, including waste wood and other natural materials

**biosphere** all the organisms on Earth and their habitat

**blast furnace** furnace for extracting iron from iron ore

**boiling** change of state from liquid to gas that happens at the boiling point of a substance

**brass** alloy of copper and zinc

**brittle** easily cracked or broken by hitting or bending

**bronze** alloy of copper and tin

**calcium carbonate** compound with chemical formula $CaCO_3$ – main component of limestone

**cancer** life-threatening condition where body cells divide uncontrollably

**carbon dioxide** one of the gasses emitted from burning fossil fuels that contributes to global warming

**carnivore** an animal that eats other animals

**carriers** someone who carries an abnormal gene but does not themselves have the disease

**cast iron** iron containing 3–4% carbon – used to make objects by casting

**cathode** negative electrode

**cement** substance made by heating limestone with clay – when mixed with water it sets hard like stone

**central nervous system (CNS)** collectively the brain and spinal cord

**charge** particles or objects can be positively or negatively electrically charged, or neutral: similar charges repel each other, opposite charges attract

**chemical bond** attractive force between atoms that holds them together (may be covalent or ionic)

**chemical equation** line of chemical formulae showing what reacts and what is produced during a chemical reaction

**chemical reaction** process in which one or more substances are changed into other substances – chemical reactions involve rearranging atoms and energy changes

**chlorophyll** pigment that is used in photosynthesis and gives green plants their colour

**chromosome** thread-like structure in the cell nucleus that carries genetic information

**climate change** changes in seasonal weather patterns that occur because the average temperature of Earth's surface is increasing through global warming

**clone** group of genetically identical organisms

**coal** solid fossil fuel formed from plant material – composed mainly of carbon

**combustion** process where substances react with oxygen, releasing heat

**competition** result of more than one organism needing the same resource, which is in short supply

**composite** a material that combines two or more materials in a mixture

**compost** partly rotted organic material, used to improve soil for growing plants

**compound** substance composed of two or more elements that are chemically joined together, for example $H_2O$

**compression** region of a longitudinal wave where the vibrating particles are squashed together more than usual

**concrete** mixture of cement, sand, aggregate and water

**conduction (electrical)** transfer of energy when an electrical current passes through a material

**conduction (thermal)** transfer of energy through a substance when it is heated

**conductor** material that transfers energy easily

**consumer** an organism that feeds on other organisms

**continental drift** movement of continents relative to each other

**continental plate** tectonic plate carrying large landmass, though not necessarily a whole continent

**convection** heat transfer in a liquid or gas  – when particles in a warmer region gain energy and move into cooler regions, carrying this energy with them

**convection current** when particles in a liquid or gas gain energy from a warmer region and move into a cooler region, being replaced by cooler liquid or gas

**core (of Earth)** layer in centre of Earth, consisting of a solid inner core and molten outer core

**corrosion** deterioration due to chemical action, as in the oxidation of iron in the presence of water

**cosmic microwave background radiation (CMBR)** microwave radiation coming very faintly from all directions in space

**covalent bond** bond between atoms in which some of the electrons are shared

**cracking** oil refinery process that breaks down large hydrocarbon molecules into smaller ones

**crust** surface layer of Earth made of tectonic plates

**culture** a population of microorganisms, grown on a nutrient medium

**current (electric)** flow of electrons in an electric circuit

**cystic fibrosis** genetic condition where the lungs become clogged with mucus

**decay (biological)** the breakdown of organic material by microorganisms

**decomposer** organism that feeds on dead plants, algae and animals

**diabetes** disease in which the body cannot control its blood sugar level

**digestion** process by which the body breaks down large food molecules into smaller ones

**displayed formula** formula of a chemical showing all the atoms and all the bonds

**distillation** process for separating liquids by boiling them, then condensing the vapours

**DNA** polymer molecule found in the nucleus of all body cells – its sequence determines genetic characteristics, such as eye colour

**Doppler effect** change in wavelength and frequency of a wave that an observer notices if the wave source is moving towards them or away from them

**double covalent bond** two covalent bonds between the same pair of atoms – each atom shares two of its own electrons plus two from the other atom

**drug** a chemical that changes the chemical processes in the body

**earthquake** shaking and vibration at the surface of the Earth resulting from underground movement or from volcanic activity

**effector** part of the body that responds to a stimulus

**efficiency** a measure of how effectively an organism or appliance transfers the energy that flows in into useful effects

**egg cell** female gamete

**electrical power** a measure of the amount of energy supplied each second

**electricity generator** device for generating electricity

**electrode** solid electrical conductors through which the current passes into and out of the liquid during electrolysis – and at which the electrolysis reactions take place

**electrolysis** decomposing an ionic compound by passing an electric current through it while molten or in solution

**electrolyte** liquid that conducts electricity in an electrolysis cell

**electromagnetic (EM) radiation** energy transferred as electromagnetic waves

**electromagnetic spectrum** electromagnetic waves ordered according to wavelength and frequency – ranging from radio waves to gamma rays

**electromagnetic waves** a group of waves that transfer energy – they travel at the speed of light and can travel through a vacuum

**electron** small particle within an atom that orbits the nucleus (it has a negative charge)

**electronic configuration** the arrangement of electrons in shells, or energy levels, in an atom

**electroplating** process of coating a surface with metal

**element** substance made out of only one type of atom

**embryo** a very young organism, which began as a zygote and will become a fetus

**emit** an object emits energy when energy is transferred away from the object as radiation

**energy** the ability to 'do work'

**energy input** the energy transferred to an organism or appliance from elsewhere

**energy levels** the arrangement of electrons in atoms (shells)

**energy output** the energy transferred away from an organism or appliance – it can be either useful or wasted

**environment** an organism's surroundings

**enzyme** biological catalyst that increases the speed of a chemical reaction

**ethanol** an alcohol that can be made from sugar and used as a fuel

**evaporation** change of state where a substance changes from liquid to gas at a temperature below its boiling point

**evolution** a change in a species over time

**extremophile** an organism that can live in conditions where a particular factor, such as temperature or pH, is outside the range that most other organisms can tolerate

**fermentation** process in which yeast converts sugar into ethanol (alcohol)

**fertilisation** fusion of the nuclei of a male and a female gamete

**fertiliser** chemical put on soil to increase soil fertility and allow better growth of crop plants

**finite resource** material of which there is only a limited amount – once used it cannot be replaced

**flammable** catches fire and burns easily

**food chain** flow diagram showing how energy is passed from one organism to another

**formula (for a chemical compound)** group of chemical symbols and numbers, showing which elements, and how many atoms of each, a compound is made of

**formulation** recipe for and form (tablet or liquid) that a medicine takes

**fossil fuel** fuel such as coal, oil or natural gas, formed millions of years ago from dead plants and animals

**fractional distillation** process that separates the hydrocarbons in crude oil according to size of molecules

**fractionating column** tall tower in which fractional distillation is carried out at an oil refinery

**fractions** the different substances collected during fractional distillation of crude oil

**frequency** the number of waves passing a set point per second

**fuel** a material that is burned for the purpose of generating heat

**fungus (pl. fungi)** living organisms whose cells have cell walls, but that cannot photosynthesise

**gamma ray** ionising electromagnetic radiation – radioactive and dangerous to human health

**gene** section of DNA that codes for a particular characteristic

**genetic disorder** inherited disease passed on from parents to children

**genetically modified (GM)** organism that has had genes from a different organism inserted into it

**gland** organ that secretes a useful substance

**global warming** gradual increase in the average temperature of Earth's surface

**glucagon** hormone, produced by the pancreas, that converts glucose stored in the liver to glucose

**glucose** a source of energy for cells

**glycogen** stores glucose in the body

**gravitropism** a growth response to gravity

**greenhouse gas** a gas such as carbon dioxide that reduces the amount of heat escaping from Earth into space, thereby contributing to global warming

**group** within the periodic table, the vertical columns are called groups

**habitat** area or environment where an organism or group of organisms live

**haemophilia** disease where blood lacks the ability to clot

**hazard** something that is likely to cause harm

**heart disease** blockage of blood vessels that bring blood to the heart

**heat of combustion** energy released when a hydrocarbon burns

**herbivore** an animal that eats plants

**hormones** chemicals that act on target organs in the body (hormones are made by the body in special glands)

**hot spot** area of Earth's crust heated by rising currents of magma – Hawaii is above a mid-Pacific hot spot

**hydrocarbon** compound containing only carbon and hydrogen

**hydro-degradable** plastics degrade in moist conditions

**hydroxide** ion consisting of an oxygen and a hydrogen atom (written as OH$^-$)

**immunity** you have immunity if your immune system recognises a pathogen and fights it

**indicator species** organisms that, by their presence or otherwise, indicate pollution levels

**infrared radiation** energy transferred as heat – a type of electromagnetic radiation

**insoluble** not soluble in water (forms a precipitate)

**insulator** material that transfers energy only very slowly – thermal insulators transfer heat slowly, electrical insulators do not allow an electric current to flow through them

**insulin** hormone made by the pancreas that reduces the level of glucose in the blood

**ion** atom (or groups of atoms) with a positive or negative charge, caused by losing or gaining electrons

**ionic bond** chemical bond between two ions of opposite charges

**ionise** to cause electrons to split away from their atoms (some forms of EM radiation are harmful to living cells because they cause ionisation)

**joule** unit used to measure energy

**kilowatt-hour** the energy transferred in 1 hour by an appliance with a power rating of 1 kW (sometimes called a 'unit' of electricity)

**leaching** using a chemical solution to dissolve a substance out of a rock

**lichen** small organism that consists of both a fungus and an alga

**limestone** type of rock consisting mainly of calcium carbonate

**liquid crystals** a liquid that has different optical properties when an electric current passes through it

**lithosphere** the rocky, outer section of Earth, consisting of the crust and upper part of the mantle

**longitudinal wave** a wave where the direction that the particles are vibrating is the same as the direction in which the energy is being transferred by the wave

**lymphocyte** white blood cells make antibodies to destroy them

**magma** molten rock found below Earth's surface

**mantle** semi-liquid layer of the Earth beneath the crust

**mass number** number of protons and neutrons in the nucleus of an atom

**mechanical wave** wave in which energy is transferred by particles or objects moving, such as a wave on a string or a water wave

**melting** change of state of a substance from liquid to solid

**methane** the simplest hydrocarbon, $CH_4$ – main component of natural gas

**microorganism** very small organism (living thing) that can be viewed only through a microscope – also known as a microbe

**microwaves** non-ionising radiation – used in telecommunications and in microwave ovens

**MMR** vaccine for measles, mumps and rubella

**molecular formula** formula of a chemical using symbols in the periodic table, such as $CH_4$ for methane

**molecule** two or more atoms held together by covalent chemical bonds

**molten** made liquid by keeping the temperature above the substance's melting point

**monomers** small molecules that become chemically bonded to each other to form a polymer chain

**mortar** mixture of cement, sand and water

**motor neurone** nerve cell carrying information from the central nervous system to muscles

**MRSA** a form of the bacterium *Staphylococcus aureus* that is resistant to many antibiotics

**mutation** a change in the DNA in a cell

**National Grid** network that distributes electricity from power stations across the country

**natural gas** gaseous fossil fuel formed from animals and plants that lived 100 million years ago – composed mainly of methane

**natural selection** process by which 'good' characteristics that can be passed on in genes become more common in a population over many generations ('good' characteristics mean that the organism has an advantage which makes it more likely to survive)

**nerve** group of nerve fibres

**neurone** nerve cell

**neutralisation** reaction between $H^+$ ions and $OH^-$ ions (acid and base react to makes a salt and water)

**neutron** small particle that does not have a charge – found in the nucleus of an atom

**noble gas** unreactive gas in Group 0 of the periodic table

**non-renewable** something that cannot be replaced when it has been used, such as fossil fuels and metal ores

**nuclear fission** energy released when nuclei of atoms are split

**nuclear fuel** radioactive fuel, such as uranium and plutonium, used in nuclear power stations

**nucleus** central part of an atom that contains protons and neutrons

**nutrient** substance in food that we need to eat to stay healthy, such as protein

**oceanic plate** tectonic plate under the ocean floor – it does not carry a continent

**oil (crude)** liquid fossil fuel formed from animals and plants that lived 100 million years ago

**ore** rock from which a metal is extracted, for example iron ore

**organic compound** a compound containing carbon and hydrogen, and possibly oxygen, nitrogen or other elements – living organisms are made up of organic compounds

**oxidation** process that increases the amount of oxygen in a compound – opposite of reduction

**oxo-degradable** plastics that use oxygen to degrade

**pancreas** organ which makes the hormones insulin and glucagon

**pandemic** when a disease spreads rapidly across many countries – perhaps the whole world

**Pangaea** huge landmass with all the continents joined together before they broke up and drifted apart

**pathogen** harmful organism that invades the body and causes disease

**payback time** time taken for a type of domestic insulation to 'pay for itself' – to save as much in energy bills as it cost to install

**period** horizontal row in the periodic table

**periodic table** a table of all the chemical elements based on their atomic number

**pesticide** chemical used to kill living organisms which are pests, such as rats or insects

**petroleum** liquid fossil fuel formed from animals and plants that lived 100 million years ago

**phagocytes** white blood cells that surround pathogens and digest them with enzymes

**photosynthesis** process carried out by green plants where sunlight, carbon dioxide and water are used to produce glucose and oxygen

**photosynthesis** process carried out by green plants where sunlight, carbon dioxide and water are used to produce glucose and oxygen

**phototropism** a growth response to light

**phytomining** using growing plants to absorb metal compounds from soil, burning the plants, and recovering metal from the ash

**placebo** 'dummy' treatment given to some patients, in a drug trial, that does not contain the drug being tested

**planet** large ball of gas or rock travelling around a star – for example Earth and other planets orbit our Sun

**plastics** compounds produced by polymerisation, capable of being moulded into various shapes or drawn into filaments and used as textile fibres

**pollution** presence of substances that contaminate or damage the environment

**polydactyly** having more than five digits on a hand or foot

**polyethene** plastic polymer made from ethene gas (also called polythene)

**polymer** large molecule made up of a chains of monomers

**polymerisation** chemical process that combines monomers to form a polymer: this is how polythene is formed

**power** amount of energy that something transfers each second and measured in watts (or joules per second)

**power rating** a measure of how fast an electrical appliance transfers energy supplied as an electrical current

**power station** place where electricity is generated to feed into the National Grid

**prey** animals that are eaten by a predator

**producer** organism that makes its own food from inorganic substances

**products** chemicals produced at the end of a chemical reaction

**progesterone** hormone produced by the ovary that prepares the uterus for pregnancy

**protein** molecule made up of amino acids (found in food of animal origin and also in plants)

**proton** small positive particle found in the nucleus of an atom

**pyramid of biomass** a diagram in which boxes, drawn to scale, represent the biomass at each step in a food chain

**quarry** place where stone is dug out of the ground

**quicklime** calcium oxide

**radiation dose** amount of radiation received by a target

**radiation intensity** rate of energy arriving at a target

**radio wave** non-ionising radiation used to transmit radio and TV

**radioactive** producing energy from the break-up of atoms

**radioactive tracer** a weak source of gamma radiation injected into the body to help in diagnosis

**radiotherapy** using ionising radiation to kill cancer cells in the body

**rate of energy transfer** a measure of how quickly something moves energy from one place to another

**reactants** chemicals that are reacting together in a chemical reaction

**receptor** nerve cell that detects a stimulus

**red-shift** when lines in a spectrum are redder than expected – if an object has a red-shift, it is moving away from the observer

**reduction** process that reduces the amount of oxygen in a compound, or removes all the oxygen from it – opposite of oxidation

**reflex action** a fast, automatic response to a stimulus

**reflex arc** pathway taken by nerve impulse from receptor, through nervous system, to effector

**resistant strain (of bacteria)** a population of bacteria that is not killed by an antibiotic

**respiration** process occurring in living things in which oxygen is used to release the energy in foods

**risk** the likelihood of a hazard causing harm

**Sankey diagram** diagram showing how the energy supplied to something is transferred into 'useful' or 'wasted' energy

**secretion** production and release of a useful substance

**selective breeding** individuals showing the required characteristic are chosen for breeding

**sensory neurone** nerve cell carrying information from receptors to the central nervous system

**shape memory alloy** alloy that 'remembers' its original shape and returns to it when heated

**shells** electrons are arranged in shells (or orbits) around the nucleus of an atom

**sickle-cell anaemia** hereditary disease where red blood cells have a curved shape

**smart material** material which changes with a stimulus

**smelting** extracting metal from an ore by reduction with carbon – heating the ore and carbon in a furnace

**speed of light** speed at which electromagnetic radiation travels through a vacuum – 300 000 000 metres per second

**sperm** male sex cell of an animal

**spinal cord** transmits information between the brain and the rest of the body

**stainless steel** steel alloy containing chromium and nickel to resist corrosion

**state symbol** symbol used in equations to show whether something is solid, liquid, gas or in solution in water

**steel** alloy of iron and steel, with other metals added depending on its intended use

**step-down transformer** transformer that changes alternating current to a lower voltage

**step-up transformer** transformer that changes alternating current to a higher voltage

**stimulus** a change in the environment that is detected by a receptor

**structural formula** formula showing the arrangement of atoms from which the molecule is made

**subduction zone** area of ocean floor where an oceanic plate is sinking beneath a continental plate

**sugar** sweet-tasting compound of carbon, hydrogen and oxygen such as glucose or sucrose

**sulfur dioxide** poisonous, acidic gas formed when sulfur or a sulfur compound is burned

**surface area** a measure of the area of an object that is in direct contact with its surroundings

**sweat** liquid secreted onto the surface skin that has a cooling effect as it evaporates

**symbol (for an element)** one or two letters used to represent a chemical element, for example C for carbon or Na for sodium

**synapse** gap between two neurones

**synthetic** made by people

**target organ** the part of the body affected by a hormone

**tectonic plate** section of Earth's crust that floats on the mantle and slowly moves across the surface

**thalidomide** a drug that was originally prescribed to pregnant women but was found to cause deformities in fetuses

**thermoplastic** polymer that can be softened and shaped repeatedly

**thermoreceptor** receptor that detects change in temperature

**thermoregulation** how the body keeps a constant internal temperature

**thermosetting** polymer that can be softened and shaped only once

**tissue** group of cells that work together and carry out a similar task, such as lung tissue

**toxin** poisonous substance (pathogens make toxins that make us feel ill)

**transfer (energy)** energy transfers occur when energy moves from one place to another, or when there is a change in the way in which it is observed

**transverse wave** wave in which the vibration of particles is at right angles to the direction in which the wave transfers energy

**turbine** device for generating electricity – the turbine has coils of wire that rotate in a magnetic field to generate electricity

**ultraviolet radiation** electromagnetic radiation that can damage human skin

**U-value** a measure of how easily energy is transferred through a material as heat

**vaccine** killed microorganisms, or living but weakened microorganisms, that are given to produce immunity to a particular disease

**vacuum** a space where there are no particles of any kind

**vaporise** change from liquid to gas (vapour)

**variation** differences between individuals belonging to the same species

**virus** very small infectious organism that reproduces within the cells of living organisms and often causes disease

**volcano** landform (often a mountain) where molten rock erupts onto the surface of the planet

**voltage** a measure of the energy carried by an electric current (the old name for potential difference)

**wasted energy** energy that is transferred by an organism or appliance in ways that are not wanted, or useful

**watt** unit of energy transfer – one watt is a rate of energy transfer of one joule per second

**wavelength** distance between two wave peaks

**X-rays** ionising electromagnetic radiation – used in X-ray photography to generate pictures of bones

# Index

# Acknowledgements

The publishers wish to thank the following for permission to reproduce photographs. Every effort has been made to trace copyright holders and to obtain their permission for use of copyright materials. The publishers will gladly receive any information enabling them to rectify any error or omission at the first opportunity.

Cover D. Roberts/Science Photo Library, p. 8t James Thew/Shutterstock, p. 8u Chapelle/Shutterstock, p. 8l Hiob/iStockphoto, p. 9t NASA, p. 9u Johann Helgason/Shutterstock, p. 9l Paul Rapson/Science Photo Library, p. 9b Juburg/Shutterstock, p. 10t NASA/JPL/Cornell University, p. 10c NASA, p. 10b NASA, p. 11 NASA, p. 12t Wikimedia Commons, p. 12b nfsphoto/Shutterstock, p. 13 NASA, p. 14 NASA, p. 16t Josemaria Toscano/iStockphoto, p. 16l Ragnarock/Shutterstock, p. 17 picturepartners/Shutterstock, p. 18 Alex Coppel/Newspix/Rex Features, p. 19 Bartosz Hadyniak/iStockphoto, p. 20 Johann Helgason/Shutterstock, p. 22t Dr. Peter Siver, Visuals Unlimited/Science Photo Library, p. 22b Scorpp/Shutterstock, p. 23t Leene/Shutterstock, p. 23c Dudarev Mikhail/Shutterstock, p. 23b Sheila Terry/Science Photo Library, p. 24 Alleyn Plowright/iStockphoto, p. 28 Danylchenko Iaroslav/Shutterstock, p. 29t enviromantic/iStockphoto, p. 29b Andrew Lambert Photography/Science Photo Library, p. 30 Wikimedia Commons, p. 32t Alexey Demidov/Shutterstock, p. 32c Charles D. Winters/Science Photo Library, p. 32b Charles D. Winters/Science Photo Library, p. 34t Robert Gubbins/Shutterstock, p. 34b John Darch/Wikimedia Commons, p. 36t Paul Rapson/Science Photo Library, p. 36b photobank.kiev.ua/Shutterstock, p. 38t Philippe-Jacques de Loutherbourg/Wikimedia Commons, p. 38cl Denis Selivanov/Shutterstock, p. 38cr inxti/Shutterstock, p. 38b Ocean/Corbis, p. 40t SueC/Shutterstock, p. 40b Walter Baxter/Geograph.co.uk, p. 41 Zybr/Shutterstock, p. 42t Wikimedia Commons, p. 42b Cultura Creative/Alamy, p. 43 Martyn F. Chillmaid/Science Photo Library, p. 44t CCI Archives/Science Photo Library, p. 44c Sotiris Zaferis/Science Photo Library, p. 44b Juburg/Shutterstock, p. 45t Martyn F Chillmaid/Photolibrary, p. 45b Bluerain/Shutterstock, p. 46 Lighttrace Studio/Alamy, p. 47 George Cairns/iStockphoto, p. 50t Gelpi/Shutterstock, p. 50c Viorel Sima/Shutterstock, p. 50b loriklaszlo/Shutterstock, p. 51t Nik Niklz/Shutterstock, p. 51c Elena Elisseeva/Shutterstock, p. 51b Monty Rakusen/cultura/Corbis, p. 52t Jarrod Boord/Shutterstock, p. 52b Snowshill/Shutterstock, p. 53 imagebroker/Alamy, p. 54t Penn State University/Science Photo Library, p. 54l Nik Niklz/Shutterstock, p. 54r Carol Gregory/Shutterstock, p. 55t Seleznev Oleg/Shutterstock, p. 55b Wolfgang Kaehler/Corbis, p. 56tl Eric Isselée/Shutterstock, p. 56tr Maxim Kulko/Shutterstock, p. 56b Yuri Arcurs/Shutterstock, p. 58t Rena Schild/Shutterstock, p. 58b Web Picture Blog/Shutterstock, p. 59t Cathy Melloan/Alamy, p. 59c marema/Shutterstock, p. 59b filmlessphotos/Shutterstock, p. 60t PRIMA/Shutterstock, p. 60b Mares Lucian/Shutterstock, p. 61 Robert HENNO/Alamy, p. 64t David Hosking/Alamy, p. 64l Fesus Robert/Shutterstock, p. 64cl Rob Stark/Shutterstock, p. 64cr sarah2/Shutterstock, p. 64r Stephen Mcsweeny/Shutterstock, p. 65t blickwinkel/Alamy, p. 65b Roca/Shutterstock, p. 66 Sea World of California/Corbis, p. 67 Adriatic2Alps Photography Tours/iStockphoto, p. 68t British Museum/Munoz-Yague/Science Photo Library, p. 68b Georgy Markov/Shutterstock, p. 70t Monty Rakusen/cultura/Corbis, p. 70b Elena Elisseeva/Shutterstock, p. 72 Mogens Trolle/Shutterstock, p. 82t iDesign/Shutterstock, p. 82c Lauri Patterson/iStockphoto, p. 82b Monkey Business Images/Shutterstock, p. 83t Alexander Raths/Shutterstock, p. 83cl Andrew Lambert Photography/Science Photo Library, p. 83cr Martyn F. Chillmaid/Science Photo Library, p. 83b Andrey Arkusha/Shutterstock, p. 84 Noam Armonn/Shutterstock, p. 85 Lobke Peers/Shutterstock, p. 86 DenisKlimov /Shutterstock, p. 87 David Reed/Corbis, p. 88 andras_csontos/Shutterstock, p. 90 Rob Byron/Shutterstock, p. 91 Alexander Raths/Shutterstock, p. 92 Tom Plesnik/Shutterstock, p. 93 Craig Eisenberg/Alamy, p. 96 Noam Armonn/Shutterstock, p. 98 Henrik Larsson/Shutterstock, p. 99b parema/iStockphoto, p. 100 David Lee/Alamy, p. 101r Martyn F. Chillmaid/Science Photo Library, p. 101l Andrew Lambert Photography/Science Photo Library, p. 102t Purestock/Alamy, p. 102b Science Source/Science Photo Library, p. 103 razorpix/Alamy, p. 104t Jakub Pavlinec/Shutterstock, p. 104b Dr Tim Evans/Science Photo Library , p. 105 Jim West/Alamy, p. 106 Benjamin Albiach Galan/Shutterstock, p. 107 CNRI/Science Photo Library, p. 108t holbox/Shutterstock, p. 108c Mandy Godbehear/Shutterstock, p. 108b Andrey Arkusha/Shutterstock, p. 109 OJO Images Ltd/Alamy, p. 110t Shvaygert Ekaterina/Shutterstock, p. 110c Will & Deni McIntyre/Science Photo Library, p. 110b Mark Burnett/Science Photo Library, p. 112 Simon Fraser/RVI, Newcastle-upon-Tyne/Science Photo Library, p. 113t CNRI/Science Photo Library, p. 113b Omikron/Science Photo Library, p. 118t Lance Bellers/Shutterstock, p. 118c Konstantin Sutyagin/Shutterstock, p. 118b Chris Parypa/Alamy, p. 119t Janis Lacis/Shutterstock, p. 119c Justin Kase zsixz/Alamy, p. 119b demarcomedia/Shutterstock, p. 120t Goran Bogicevic/Shutterstock, p. 120c jorisvo/Shutterstock, p. 120b WDG Photo/Shutterstock, p. 121 Martyn F. Chillmaid/Science Photo Library, p. 122t The Print Collector/Alamy, p. 122c johnnyscriv/iStockphoto, p. 122b Janis Lacis/Shutterstock, p. 123 Gontar/Shutterstock, p. 124t Chris leachman/Shutterstock, p. 124b clearviewstock/Shutterstock, p. 125tr Lance Bellers/iStockphoto, p. 125t jan

kranendonk/Shutterstock, p. 125c indianstockimages/Shutterstock, p. 125b Charle Stirling/Alamy, p. 126 David Hughes/Shutterstock, p. 127 marilyn barbone Shutterstock, p. 128t qaphotos.com/Alamy, p. 128b TLF Design/Alamy, p. 129 Nicola Gavin/Shutterstock, p. 129c fotosav/Shutterstock, p. 129b courtesy of Shangy Teaching Instrument Factory, p. 130 Vladimir Menkov/Wikimedia Commons, p. 131 Levent Konuk/Shutterstock, p. 131b LesPalenik/Shutterstock, p. 134t Justin Kas zsixz/Alamy, p. 134b Ian Maybury/Alamy, p. 136t photobank.kiev.ua/Shutterstock, p. 136c Danny E Hooks/Shutterstock, p. 136b Karol Kozlowski/Shutterstock, p. 138 Darren Baker/Shutterstock, p. 138b Igor Grochev/Shutterstock, p. 139t Yegor Korzh Shutterstock, p. 139b Idealink Photography/Alamy, p. 140 Yves Grau/iStockphoto, p. 141 ilbusca/iStockphoto, p. 142t Karl-Friedrich Hohl/iStockphoto, p. 142b Michae DeGasperis/Shutterstock, p. 143t Apple's Eyes Studio/Shutterstock, p. 143 demarcomedia/Shutterstock, p. 146t Media3d/Shutterstock, p. 146b Theo Gottwald Alamy, p. 147t stephen mulcahey/Shutterstock, p. 147b James Doss/Shutterstock, p 148t courtesy of Husqvarna, p. 148b Agita Leimane/Shutterstock, p. 149 stephe mulcahey/Shutterstock, p. 150 Paul M Thompson/Alamy, p. 151 Joey Boylan iStockphoto, p. 152 Morgan Lane Photography/Shutterstock, p. 153 Realimage Alamy, p. 154t thesuperph/iStockphoto, p. 154b jiri jura/Shutterstock, p. 155t Eri Broder Van Dyke/Shutterstock, p. 155b Benis Arapovic/Shutterstock, p. 156 courtesy of Smartwater Technology, p. 158t Mark Kostich/iStockphoto, p. 158b Ted Kinsman Science Photo Library, p. 159t Terry Williams/Rex Features, p. 159c Philippe Psaila Science Photo Library, p. 159b Paul Gooney/Alamy, p. 160 Baris Simsek/iStockphoto p. 170t alex saberi/Shutterstock, p. 170c Goodluz/Shutterstock, p. 170b valdis torms Shutterstock, p. 171t Lusoimages/Shutterstock, p. 171c Juergen Berger/Science Photo Library, p. 171b Mark Kostich/iStockphoto, p. 172t Jason Stitt/Shutterstock, p. 172 Lusoimages/Shutterstock, p. 172r Picsfive/Shutterstock, p. 173 Blaj Gabrie Shutterstock, p. 174t Westend61 GmbH/Alamy, p. 174b Anthony Hall/Shutterstock, p 176t Island Effects/iStockphoto, p. 176b iofoto/Shutterstock, p. 177t Hulton Archive Getty Images, p. 177b marilyn barbone/Shutterstock, p. 178 T.W. van Urk/Shutterstock. p. 179t Jorge Cubells Biela/Shutterstock, p. 179b Rena Schild/Shutterstock, p. 180 World History Archive/Alamy, p. 181 Tramper/Shutterstock, p. 184t NIBSC/Science Photo Library, p. 184b Dr Jeremy Burgess/Science Photo Library, p. 185 NIBSC Science Photo Library , p. 186t Dr. P Marazzi/Science Photo Library, p. 186b Nationa Cancer Institute/Science Photo Library, p. 187 Juergen Berger/Science Photo Library p. 188t PHIL/Barbara Rice, p. 188b Dmitry Naumov/Shutterstock, p. 190 C. Powell, P.Fowler & D. Perkins/Science Photo Library, p. 191 Dmitry Rukhlenko/Shutterstock, p. 192t Gonul Kokal/Shutterstock, p. 192b Health Protection Agency/Science Photo Library, p. 193 Mark Kostich/Shutterstock, p. 194t jannoon028/Shutterstock, p. 194b Doncaster and Bassetlaw Hospital/Science Photo Library, p. 195 ISM/Science Photo Library , p. 196t The Print Collector/Alamy, p. 196b Steve Gschmeissner/Science Photo Library/Alamy, p. 198 Mary Evans Picture Library/Alamy, p. 202t Scot Rothstein Shutterstock, p. 202c great_photos/Shutterstock, p. 202b Adrian T Sumner/Science Photo Library, p. 203t mustafa deliormanli/iStockphoto, p. 203c Max Earey/Shutterstock, p. 203b Jane Rix/Shutterstock, p. 204t mustafa deliormanli/iStockphoto, p. 204c Alexandr Makarov/Shutterstock, p. 204b Dr. P Marazzi/Science Photo Library, p. 206 Rechitan Sorin/Shutterstock, p. 207 Trevor Clifford Photography/Science Photo Library, p. 208t Manfred Kage/Science Photo Library, p. 208b sgame/Shutterstock, p. 209t Max Earey/Shutterstock, p. 209b Reuters/Corbis, p. 210t Science Photo Library/Alamy, p. 210b Phil Degginger/Alamy, p. 211 Peter Menzel/Science Photo Library/Alamy, p. 214t IDAL/Shutterstock, p. 214b Jane Rix/Shutterstock, p. 215t FotoJagodka/Shutterstock, p. 215b Wikimedia Commons, p. 216 R.L. Brinster, Peter Arnold Inc./Science Photo Library, p. 217 chungking/Shutterstock, p. 220l Kodda/Shutterstock, p. 220r Martin Shields/Alamy, p. 220b Can Balcioglu/Shutterstock, p. 221t ilFede/Shutterstock, p. 221c courtesy of NIA, p. 221b Coyote-Photography.co. uk/Alamy, p. 222t Robert Brook/Science Photo Library, p. 222b dcwcreations/Shutterstock, p. 223t Satin/Shutterstock, p. 223b fritz16/Shutterstock, p. 224t wrangler/Shutterstock, p. 224c ilFede/Shutterstock, p. 224r Yuriy Boyko/Shutterstock, p. 224l svic/Shutterstock, p. 225t cubephoto/Shutterstock, p. 225b Martin Shields/Science Photo Library, p. 226t stephen bond/Alamy, p. 226b Blazej Lyjak/Shutterstock, p. 228 PeskyMonkey/iStockphoto, p. 230t MARKABOND/Shutterstock, p. 230b Peter Alvey/Alamy, p. 231 courtesy of NIA, p. 232 Olena Mykhaylova/Shutterstock, p. 234t Bubbles Photolibrary/Alamy, p. 234b Peter Mukherjee/iStockphoto, p. 235 Eye of Science/Science Photo Library, p. 236 Thomas M Perkins/Shutterstock, p. 237 Coyote-Photography.co.uk/Alamy, p. 238t sommthink/Shutterstock, p. 238b mark phillips/Alamy, p. 239 Leslie Garland Picture Library/Alamy, p. 240 VR Photos/Shutterstock, p. 254t Andrew Lambert Photography/Science Photo Library, p. 254c Pedro Salaverría/Shutterstock, p. 254b Shawn Hempel/Shutterstock, p. 260 Martyn F. Chillmaid/Science Photo Library, p. 274-275 Yuri Arcurs/Shutterstock, p. 274bl Valua Vitaly/Shutterstock, p. 274br Martina Ebel/Shutterstock, p. 275tr Photoroller/Shutterstock, p. 276-277 Pixel 4 Images/Shutterstock.